Land and Farm Resources in Forestry

Land and Farm Resources in Forestry

Musharraf Hussain

RANDOM

RANDOM PUBLICATIONS
NEW DELHI (INDIA)

Land and Farm Resources in Forestry

978-93-5111-164-1

Published in 2013 in India by

RANDOM PUBLICATIONS

4376-A/4B, Gali Murari Lal, Ansari Road
New Delhi-110 002
Phone : +9111-43580356, 011-23289044
e-mail: randompublications@hotmail.com

Reprinted 2018

Type Setting by : Friends Media, Delhi-110089
Digitally Printed at : Replika Press Pvt. Ltd.

Preface

These forest lands provide the direct resource base for the forest industry and an indirect base for related industries. A majority of the water resources of the County originate in the forested areas of the County. These forests also provide abundant wildlife habitat and areas that are widely used for outdoor recreation. Forest lands serve a multitude of functions. The unique scenic and environmental qualities of forest lands make them attractive for recreational activities such as camping, hiking, fishing, hunting, water sports, etc. These activities, in addition to providing an important social benefit, also contribute significantly to the economy.

To conserve forest lands and mixed farm/timber lands by maintaining the forest land base and the mixed forest and agricultural base and to protect the county's forest and farm economies by making possible economically efficient forest and agricultural practices that assure the continuous growing and harvesting of forest tree species and agricultural products as the leading use on forest land and mixed farm/timber land consistent with sound management of soil, air, water and fish and wildlife resources and to provide for recreational opportunities.

Forest farming, the cultivation of under story plants and fungi with economic value, may be a way for small-scale forest landowners to realize greater benefits. The forests are host of a tremendous diversity of economically important plants. Eighty percent of the land base of the Appalachian forest region is in private ownership, and the majority of this is in family forests. These holdings are typically small in size, and maintained for reasons other than timber production. A great diversity of native plant species are harvested for their value in the medicinal, culinary, craft, and other product markets. Most of the plants used for their non-timber values are harvested from the wild, with little or no consideration for the long-term impact on the natural resources.

This book deals with issues of efficient utilization of natural resources, land and trees particularly in dry land areas.

—Author

Contents

1

Introduction

LAND AND FARM RESOURCES

Land is among the most important resources used by U.S. agriculture—and its chief finite resource. Decisions on land use are made daily by operators of the Nation's 2.1 million farms, contingent on market conditions, public policies, and farm/ranch characteristics. This chapter examines land use, farm real estate values, and the relationship between land ownership, farm size, and other characteristics. Among the findings:

- Land continues to shift between agriculture and other uses. Cropland has declined over recent decades, but losses do not threaten the Nation's capacity to produce food and fibre.
- Real estate is a major asset for farm households. Values have been rising steadily since 1987, but the inflation-adjusted (real) value of U.S. farm real estate is still below its 1982 peak.
- Small family farms account for most land owned by farms.

LAND USE: ENVIRONMENT AND THE SUSTAINABILITY

Land-use changes can affect the environment and the sustainability of production. Because impacts on the environment—including erosion, water quality, and wildlife habitat—are typically not reflected in private profit calculations, land-use choices that are optimal for an individual may not be optimal for society. This difference suggests the possibility of public policies that more closely align land-use decisions with social objectives. The allocation of a fixed land base among competing uses is determined by the relative returns to the different uses, which vary according to land quality and location. A landowner seeking to maximize profits will allocate a land parcel to the use that yields the highest expected economic return, after the costs of conversion. As relative returns change along with market conditions, technological advancements, or government policies, land-use patterns tend to adjust accordingly. Land-use change is dynamic. With the exception of urban land, changes occur to and from major land uses. For example, 44 million acres left the cropland and pasture category from 1992 to 1997 while 21 million acres shifted into the category, resulting in a net loss of 23 million acres.

MAJOR LAND USES

This series contains acreage estimates of major uses by region and State, coinciding with each census of agriculture from 1945 through 2002. Because Alaska and Hawaii have very little crop area, we focus on the contiguous 48 States. The total land area of the 48 contiguous States is approximately 1.9 billion acres, with an additional 365 million acres in Alaska and a little over 4 million acres in Hawai.

Grassland pasture and range, the largest use of land, accounted for 584 million acres (31 percent) of the 48 States in 2002. This compares with 636 million acres in the mid-1960s. Due to improvements in the forage quality and productivity of grazing lands, less pasture and range is needed to sustain grazing herds. The inventory of domestic animals, particularly sheep, has also been declining in recent years, further reducing pasture/range demand.

Forest-use Land

Table: Major uses of land, United States, 2002[1]

Land use	*48 States*	*US*	*48 States*	*US*
	Million acres		*% of total*	
Cropland[2]	441	442	23.3	19.5
Grassland pasture and range	584	587	30.8	25.9
Forest-use land[3]	559	651	29.5	28.8
Special uses[4]	153	297	8.1	13.1
Urban	59	60	3.1	2.6
Miscellaneous other land	97	228	5.1	10.1
Total land area[5]	1,894	2,264	100.0	100.0

[1]See Major Land Uses for estimates of major uses by region and State, coinciding with each census of agriculture, from 1945 through 2002.

[2]All land in the crop rotation (used for crops, used for pasture, idle cropland). Includes about 34 million acres idled under the Conservation Reserve Programme.

[3]Total forest land as classified by the U.S. Forest Service minus an estimated 98 million acres of forested land used for parks, wildlife areas, and other special uses.

[4]Rural transportation areas, land used primarily for recreation and wildlife purposes, various public installations and facilities, farmsteads, and farm roads/lanes. Excludes urban land in contrast to Major Land Uses, Aggregate Data.

[5]Distributions by major use may not add to totals due to rounding.

Sources: USDA/ERS based primarily on reports and records of the Census Bureau and Federal, State, and local land management and conservation agencies.

Forest-use land, the second largest major use, declined from about 32 percent of total land in 1945 to about 30 percent in 2002. A broader category, all land with forest cover, comprised 33 percent of the land base in 2002. While forest-use land increased 1 percent between 1997 and 2002, it declined from 612 million acres in 1964 to 559 million acres in 2002. Much forest-covered land is in "special uses" that prohibit forestry uses such as timber production. Forested land in these special uses increased from 23 million acres in 1945 to about 98 million acres in 2002.

Cropland comprises the third largest use of land, covering 23 percent of the contiguous States in 2002. Since 1945, cropland ranged from a high of 478 million acres in 1949 to a low of 441 million acres in 2002. Total cropland has trended downward since the late 1960s, and decreased by 13 million acres (3 percent) from 1997 to 2002. The total cropland base includes cropland used for crops, cropland used for pasture, and cropland idled. These components vary more than total cropland. Since 1945, the amount of cropland used for crops has ranged fromas much as 383 million acres in 1949 and 1982 to a minimum of 331 million acres in 1987. Total acreage used for crops exhibited two major cycles between 1945 and 1987, with cropland moving from idle to crop use and back again. Cropland used for crops increased from 331 to 349 million acres over 1987-97, and then declined to 340 million acres in 2002, about 5 percent below the average acreage for 1910-97. Since 1945, cropland used for pasture varied from 47 million acres in 1945 to 88 million acres in 1969.

LAND USED FOR RECREATION AND WILDLIFE AREAS

Special uses include rural transportation; rural parks and wildlife; defence and industrial uses; and farmstead, farm roads/lanes, and other onfarm uses. These special uses increased from 85 million acres (4 percent of the land area of the contiguous 48 States) in 1945 to 153 million acres (8 percent) in 2002. Land in transportation uses (highways and roads, railroads, and airports in rural areas) increased by 4 million acres (17 percent) between 1945 and 1982. Transportation uses declined by about 0.5 million acres from 1982 to 1992 due to the abandonment of railroad facilities and rural roads, and the classification of some transportation uses as urban areas.

Land used for recreation and wildlife areas (Federal and State parks, wilderness areas, and wildlife refuges) expanded 344 percent from 1945 to 2002 (an increase of 78 million acres). The increase came mostly from conversion of Federal lands, previously in forest and grassland pasture and range. Land in defence and industrial uses declined by 10 million acres (40 percent) from 1945 to 2002. Farmsteads, farm roads, and other farm uses declined by 4 million acres (29 percent) between 1945 and 1997. This decline reflects trends toward fewer farms and larger, more consolidated farms, as well as an increasing tendency for farm households to live off the farm. In response to expanding U.S. population, land in urban uses—including homes, schools, office buildings, shopping sites, and other commercial/industrial uses—increased from 15 million acres in 1945 to 25 million acres in 1960, 47 million acres in 1980, and 59 million acres in 2002. While the U.S. population nearly doubled, the amount of land urbanized quadrupled. However, urban uses still comprise only 3 percent of the total land area of the contiguous States.

Miscellaneous other land uses decreased from 1945 to 1964, and have since trended upward, showing a 54-percent increase from 1964 to 2002 reflecting improved data and reclassification of grazing and forest lands. These uses include marshes and open swamps not included in other major land uses, bare rock areas, deserts, some rural residential areas, and other uses not inventoried. Wetlands are defined by soil and hydrological characteristics and may occur on land in many different uses.

REGIONAL CHANGES IN LAND USE

While land in every use occurs in all 10 regions of the contiguous States, some uses are more concentrated in some regions than in others. Regions with the largest cropland acreage are the Northern Plains, Corn Belt, and Southern Plains. Grassland pasture and range is concentrated in the Mountain and Southern Plains regions. Acreage in forest use, special and miscellaneous other uses is highest in the Mountain region.

The Northeast, Appalachian, Southeast, Delta States, and Lake States regions lost cropland between 1945 and 2002. The largest increases occurred in the Northern Plains and Mountain regions, with smaller increases in the Corn Belt, Southern Plains, and Pacific regions. Western increases may have resulted in part from federally subsidized irrigation water.

Nine of the 10 regions lost grassland pasture and range between 1945 and 2002. While grassland pasture and range increased 11 million acres (10 percent) in the Southern Plains, the Northeast region lost about 70 percent of its grassland pasture and range, and the Appalachian and Lake States regions lost more than 50 percent. The Northeast and Appalachian regions saw the reforestation of grassland, loss of some grassland to urbanization, and concentration of the dairy industry. Decreases in the Corn Belt, Northern Plains, and Mountain regions were likely associated with the conversion of some grassland pasture and range to cropland.

Cropland Use and Federal Programmes

While total cropland acreage has varied up and down and generally declined since 1969, greater shifts have occurred between cropland used for crops and cropland idled, mostly because of Federal programmes. Cropland used for pasture has exhibited less variation in acreage than cropland idled. Most cropland used for crops is harvested, but typically 2-3 percent experiences crop failure and 5-10 percent is cultivated summer fallow. In 2002, farmers harvested one or more crops on an estimated 307 million acres of cropland, down 4 percent from 1997. About 8 million acres of the total harvested were double-cropped. When double-cropped land is counted twice, total acres harvested rise to 315 million acres.

Table: Major uses of cropland, 48 contiguous States, 1992-2004

Cropland	*1992*	*1997*	*2002*	*2004*[1]
			Million acres	
Cropland used for crops[2]	337	349	340	336
Cropland harvested[3]	305	321	307	312
Crop failure	8	7	17	9
Cultivated summer fallow	24	21	16	15
Cropland idled by all Federal programs[2]	55	33	34	35
Annual programmes	19	0	0	0
Conservation Reserve Programme	35	33	34	35
Total, specified uses[2],[4]	392	382	374	372

[1]Preliminary, subject to revision.

[2]Breakdown may not add to totals due to rounding.

[3]A double-cropped acre is counted as 1 acre.

[4]Does not include cropland pasture or idle land not in Federal programmes that is normally included in the total cropland base.

Sources: USDA/ERS, based on USDA/ERS, 1997a; USDA/NASS, 1998, 1999a, 2000a; 2004a; 2005; and unpublished data from USDA/FSA and USDA/NASS.

Cropland used for crops was at a record high of 387 million acres in 1949, when no acres were idled by Federal programmes. In 1972, cropland used for crops was near a record low of 334 million acres as Federal programmes idled 61 million acres. Cropland used for crops climbed to 387 million acres in 1981, when Federal programmes idled no cropland, and dropped to 333 million in 1983, when Federal programme set-asides reached a historic peak of 78 million acres under the Payment-in-Kind (PIK) programme. The Federal Agricultural Improvement Act (FAIR) of 1996 eliminated all Federal acreage reduction programmes other than the Conservation Reserve Programme (CRP)

Between 1983 and 2002, cropland used for crops increased overall, while total acreage idled by Federal programmes decreased to 34 million acres. From 1997 to 2002, cropland used for crops declined from 349 to 340 million acres, while acreage idled under CRP increased by about 1 million acres. Between 2002 and 2004, acreage in CRP increased by an additional 1 million acres, while cropland used for crops declined by about 4 million acres.

The 14-million-acre drop in harvested cropland between 1997 and 2002 was coincident with a decrease in cultivated summer fallow and an increase in failed acres due to widespread drought. Crop failure occurred on 17 million acres, over 5 percent of the acreage planted, in 2002. This failed acreage was the largest since 1956. The use of summer fallow has been decreasing since the late 1960s, and stood at 15 million acres in 2002, down from 42 million acres in 1969. Four crops—corn for grain, soybeans, wheat, and hay—accounted for 80 percent of all crop acres harvested in 2002. The additional 17 "principal" crops accounted for another 15 percent of harvested area. Vegetables, fruits, nuts, melons, and all other crops accounted for 4.5 percent of crop area harvested in 2002.

URBANIZATION OF AGRICULTURAL AND OTHER RURAL LAND

Cropland conversion to urban uses is largely irreversible, so it is important to know the rate of conversion and how much of the loss is replaced from other land uses. Excessive loss of cropland to urban uses could lessen the production of food and fibre and the supply of rural amenities, such as open space, watershed protection, and rural lifestyles. A variety of Federal, State, local, and private programmes address such concerns.

Although urban land constituted less than 3 percent of the U.S. land area in 2000, 79 percent of the population lived there. Even large percentage increases in urban area would amount to small decreases in rural area since it is so vast. The rate of expansion (by decade) of urban area has declined from 39 percent during the 1950s, to about 36 percent during the 1960s and the 1970s, and to 18 percent in the 1980s. According to the Census Bureau, urban area was 59 million acres in 2000, just 7 percent above the previous estimate for 1990 (DOC/BOC, 2002). However, the Census Bureau adopted a new definition of urban area for the 2000 Census, improving the precision of urban area measurement but making it more difficult to compare urban area before and after this year. If urban area for 1990 is recalculated using the 2000 definition, the Census Bureau's estimate falls to 51 million acres, implying a 15-percent increase in urban area from 1990 to 2000.

The National Resources Inventory (NRI), conducted by USDA's Natural Resources Conservation Service (NRCS) in cooperation with Iowa State University, is an alternative source for estimates of urban and rural areas. The NRI uses a consistent definition for built-up areas, though it differs from the definition used by the Census Bureau. According to the NRI, "developed land," which includes large and small urban and built-up areas as well as rural transportation land, totalled 107 million acres in 2002 in the contiguous United States. The NRI indicates that developed land increased by 14 million acres (19 percent) over 1982-92 and 21 million acres (24 percent) over 1992-2002.

Land converted to urban uses comes from several different major land uses. The NRI indicates that 20 percent of new developed land came from cropland between 1997 and 2002. Prime cropland— land that has the best combination of physical/chemical characteristics for agricultural production—is converted to developed uses at about the same rate (5 percent per year) as nonprime cropland. About 21 percent of rural non-Federal land is prime and 26 percent of crop land converted to urban uses over 1997-2001 was prime. Rural land, defined as all land that is not urban, contains rural residential land, consisting of houses and associated lots. Nonfarm rural residential area was estimated to be about 94 million acres in 2002, up from 56 million acres in 1980. The average rate of increase in rural residential land was 1.7 million acres per year from 1980 to 2002. Combining both rural and urban residential land, the total increase in residential area was about 2 million acres per year during this period.

FARM REAL ESTATE VALUES

Farmland values rose rapidly during the 1970s and early 1980s, followed by a sharp decline during 1982-87, then a slow upward trend beginning in 1987. Since 1987, average farmland values in the Nation have increased 127 percent, from $599 per acre to $1,360 in January 2004. In real or inflation-adjusted terms (GDP deflator), however, this amounts to a 53-percent gain. It was not until January 1, 1995, that the average nominal value per acre surpassed the record high of $823 set in 1982. But the January 2004 average is still 8 percent below the 1982 average on a real (or inflation-adjusted) basis.

The 7.1-percent nominal increase in the national average value of agricultural real estate during 2004 marked the 17th consecutive increase since 1987. Over the previous 4 years, in particular, farm real estate values had increased substantially in all U.S. regions. Most notable is a 42-percent increase in the Lake States, versus 25 percent for all regions combined.

Table: Farm real estate values, by farm production region, January 1 for selected years

Region *$/acre*	*1982*	*1987*	*2000*	*2001*	*2002*	*2003*	*2004*
Northeast	1,367	1,491	2,660	2,830	3,000	3,200	3,400
Lake States	1,234	707	1,560	1,700	1,870	2,010	2,220
Corn Belt	1,642	900	1,890	1,950	2,030	2,130	2,300
Northern Plains	547	331	535	556	576	594	632
Appalachian	1,083	1,004	1,990	2,120	2,250	2,370	2,500
Southeast	1,095	1,055	1,920	2,030	2,140	2,270	2,420
Delta States	1,135	757	1,270	1,330	1,390	1,490	1,550
Southern Plains	576	532	672	715	755	788	832
Mountain	325	257	448	471	500	523	550
Pacific	1,346	1,084	2,000	2,120	2,240	2,350	2,480
48 States	823	599	1,090	1,150	1,210	1,270	1,360

Source: National Agricultural Statistics Service, USDA.

The increases were widespread, with most States exhibiting increases for farm real estate, cropland, and pasture. As of January 2004, several Northeast

States continued to record the highest average per-acre values for farm real estate, with Connecticut and Rhode Island exceeding $10,000 per acre. These values reflect continued pressure from nonagricultural sources for conversion to residential or other urban-related uses. The high values in States such as California, Florida, Ohio, Illinois, Indiana, and North Carolina are the consequence of urban pressures, the production of high-value crops, or high soil fertility. The low real estate values for many States in the Northern Plains, Southern Plains, and Mountain regions can be attributed to large amounts of arid rangeland and less productive cropland. New Mexico, Wyoming, and Montana recorded the lowest average values per acre.

Grazing Fees

Grazing fees for use of pasture or rangeland are also a form of cash rent, except that payment is based on "grazing units" rather than tracts of land (acres). A grazing unit is defined on an animal-unit-month (AUM) basis, which is one cow or cow-calf pair, or seven sheep/goats, feeding for 1 month. Grazing fees on public lands administered by the Bureau of Land Management (BLM) of the U.S. Department of the Interior, and the Forest Service (FS) of the U.S. Department of Agriculture (USDA) are set by law. These fees vary annually according to a legislated formula, which links the fees to changes in the cost of production. As a result of the formula, 2005 grazing fees on public land were set at $1.79 per AUM. That marks the second consecutive year in which grazing fees were set above the statutory minimum $1.35 per AUM.

Grazing rates on privately owned nonirrigated land in 16 Western States averaged $14.30 per AUM in 2004. Rates ranged from $7 in Arizona and Oklahoma to $23 per AUM in Nebraska. Private grazing rates have trended upward since the early 1990s.

Cash Rents

Nearly a third of U.S. farmland is operated under some form of lease, according to the 2002 Census of Agriculture. The most common form of lease, the cash rental agreement, is a fixed payment negotiated before planting. Share rental agreements, by contrast, vary with the amount of product harvested. Under cash rental arrangements, the tenant bears all of the production and market-price risk; share rental arrangements divide production and market risks between tenant and landlord. Cash rents are generally considered a short-term indicator of the return to a landowner's investment. To tenants, though, cash rents are a major production expense and, like farm real estate values, have been increasing for a number of years.

Because rents reflect the income-earning capacity of the land, they vary widely across the country. Cropland rents tend to be highest in areas where higher-value crops are grown. The highest average cash rents in 2004 were reported for irrigated land in California, at $300 per acre. California produces large quantities of high-value speciality crops, vegetables, fruits, and nuts. Cropland most suitable for corn and soybean production, principally in the Midwest, also commands high rents. The highest rents for nonirrigated cropland in 2004 were reported as $126 per acre in both Illinois and Iowa.

During 2004, average cash rents for pasture varied from $37 per acre in Wisconsin to $1.70 per acre in New Mexico. States in the Appalachian, Delta, Southern Plains, and Pacific regions uniformly recorded increases from 2003.

Factors Affecting Farm Real Estate Values

Traditionally, farmland value was based on its agricultural productivity. Particularly in the more rural areas of the Nation, where farmers still account for most farmland purchases, net returns to agricultural uses are the principal determinant of farmland value. Interest rates, capital investment in farm structures, and many other factors also influence productivity and thus the agricultural value of farm real estate. But today, many factors unrelated to productivity—including urban influence, government programme payments, and rural amenities—contribute to the value of land in rural areas. In fact, these factors may be more important than productivity. High levels of direct government payments, which have occurred particularly since 1999, may have influenced farmland values in some regions.

Direct Federal Payments

An array of government policies influence the income derived from farmland, and hence its value. Federal commodity and conservation programmes are the most obvious. But also important are farm credit programmes, State and local zoning regulations, habitat and species protection laws, infrastructure development (such as roads and dams), environmental regulations, and even property and income tax policy. Previous research has shown that capitalization of expected payments increases cropland values. Also, the degree to which direct Federal payments are capitalized into cropland values depends upon the issuing programme. If direct payments are capitalized into cropland values, as many theorize and some research has demonstrated, then a reduction in payments could signal a decline in cropland values and a loss of wealth for landowners. Further, ERS estimates that the degree to which direct payments (even from the same programme) are capitalized into cropland values varies widely, with capitalization greatest in the Northern Plains. So from a policy perspective, the effect of programme changes on cropland values would vary depending on the dominant programme crop in a region.

Urban Influence

Farmland near cities has seen its value inflated by demand for conversion to nonfarm uses. As the U.S. population continues to grow and disperse, even primarily rural States such as Iowa are experiencing urban-related influences on farmland values. Commuters, who can now travel farther or even telecommute, are often willing to pay more than agricultural value in order to live in primarily rural areas. Other families develop hobby farms, second homes, or recreational structures in rural areas. In Iowa, for instance, there are now more nonfarmers living in rural areas than there are farmers. Other nonagricultural factors that may contribute value include the potential to concurrently use farmland for fee-based hunting, fee-based recreation, or wildlife viewing.

Nonfarm influences on agricultural real estate values have gained increased attention as interest in farmland preservation, suburban "sprawl," and habitat conservation has grown. Recent research indicates that nonfarm influence accounts for 25 percent of the market value of U.S. farmland. An ERS report recently addressed issues surrounding development of new houses, roads, and commercial buildings at the fringe of existing urban areas. This "sprawl" into the countryside can intersperse sometimes incompatible urban-related development with existing agriculture. Metropolitan Statistical Areas (MSAs) contain 20 percent of U.S. land area and 80 percent of U.S. population. The area also contained more than a third of all U.S. farms in 2003 and produced about a third of agricultural production value.

Other Market-Related Factors

Interest rates, particularly inflation-adjusted ones, are especially important determinants of U.S. farmland values. As proxies for the discount rate, interest rates determine the current value of expected future earnings from land: for a given pattern of future earnings, higher (lower) interest rates imply lower (higher) land values. During much of the mid-to late 1970s, real interest rates were actually negative, providing a strong incentive to borrow money. Some of the borrowed money was used to purchase rapidly appreciating farmland. Conversely, real interest rates jumped from 1981 to 1985 when nominal interest rates increased rapidly just as expectations of future inflation were diminishing. The resulting increase in the real interest rate of mortgages has been cited as a cause of the slide in farmland values in the early and mid-1980s. Inflation, lending policies of farm credit agencies and banks, and speculation also affect farmland values. And of course farmland values vary by site-specific characteristics like access to major highways, proximity to commodity and input markets, aesthetic appeal, and homesite potential.

Nonmarket Public Goods Provided by Farmland

Farmland also provides nonmonetary benefits. Until recently, these "rural amenity" benefits were supplied in such abundance that they were rarely acknowledged. But as the Nation becomes more urbanized, with the concomitant loss of farms and interspersion of urban-related activities, the decrease in those amenities has become a source of concern. The nonmonetary benefits potentially reduced or eliminated by loss of farmland and open space include recreation opportunities, aesthetic enjoyment from viewing landscapes and wildlife, environmental quality, and nostalgia related to the historic and cultural significance of rural life. It is these "rural amenity" benefits that many farmland preservation programmes seek to protect.

LAND OWNERSHIP AND FARM STRUCTURE

OWNERSHIP OF U.S. LAND

The land surface of the United States covers 2.3 billion acres. Private owners held 61 percent in 2002, the Federal Government 28 percent, State and local governments 9 percent, and Indian reservations 3 percent. Virtually all cropland is privately owned, as is three-fifths of grassland pasture and range and over half of forestland. Federal, State, and local government holdings consist primarily of forestland, rangeland, and other land. Most land in Federal ownership—largely in the West—is managed by the Department of the Interior (68 percent) and the Department of Agriculture (28 percent).

Farm operators do not own all the land used in agriculture. According to the 1999 Agricultural Economics and Land Ownership Survey (AELOS), farmers held 58 percent of the land in farms in 1999. These landowning farmers also made up 58 percent of the 3.4 million farmland owners.

Nonoperator landlords accounted for the remaining 42 percent of land in farms. Ninety-five percent of nonfarm landlords were individuals/families or partnerships. Of these unincorporated landlords, 55 percent were at least 65 years old. Many nonfarm landlords have a historic connection to farming. Among the people who have exited farming or inherited farmland since the number of farms peaked during the Great Depression, a number have retained ownership of some or all their land.

Farm Numbers, Farm Types, and Conservation Programmes

The number of farms has declined dramatically since its peak of 6.8 million in 1935, with most of the decline occurring during the 1940s, 1950s, and 1960s. The decline in farm numbers has levelled off since the 1970s. By 2002, 2.1 million farms remained. The remaining farms have a much larger average acreage, but averages mask differences among farms. Today's farms range from very small retirement and residential farms to industrialized operations with sales in the millions. Part of this diversity stems from the very low sales threshold ($1,000) necessary for an operation to qualify as a farm for statistical purposes.

One way to address the diversity of farms is to categorize them into more homogeneous groups. The farm typology developed by ERS identifies five groups of small family farms (sales less than $250,000): limited-resource, retirement, residential/lifestyle, farming-occupation/low-sales, and farming-occupation/high-sales. The typology also includes large family farms, very large family farms, and nonfamily farms. In addition, very small farms (sales less than $10,000) make up more than half of all farms. Very small farms account for a particularly large share of farms in the limited-resource (72 percent), retirement (76 percent), and residential/lifestyle (76 percent) groups. Production, however, is concentrated among larger farms; small farms account for only 27 percent of the total value of production.

The smallness of most farms has implications for conservation and the environment. An ERS study found that smaller corn farms are less likely to use conservation tillage than are larger farms. The practice is more practical for larger farms because they have more acres over which to spread the cost of new or retrofitted equipment necessary to adopt conservation tillage. Small farms whose operators are retired or farm part-time are also less likely to adopt conservation tillage, possibly because of hesitancy to change familiar production practices. Small farms, however, participate widely in the Conservation Reserve Programme (CRP) and the Wetlands Reserve Programme (WRP).

DISTRIBUTION OF CONSERVATION PROGRAMME PAYMENTS BY TYPE OF FARM

These farms harvest most of the land planted to programme commodities and therefore receive three-quarters of commodity programme payments. However, CRP and WRP—the two major conservation programmes—are targeted at particular types of land, not commodities. Since small farms own 70 percent of the land held by farms, they play a large role in natural resource and environmental policy. Retirement, residential/lifestyle, and low-sales farms account for nearly two-thirds of conservation payments and a similar share of the land farmers enrolled in the CRP and WRP. Participating farmers in each of the three groups tend to enroll large shares of their land in these programmes: 46 percent of the land operated on retirement farms, 28 percent on residential/lifestyle farms, and 23 percent on low-sales farms. In contrast, enrollment ranges from 5 to 9 percent for participating high-sales, large, and very large farms.

Because their main job is off-farm, residential/lifestyle operators are limited in the amount of time they can spend farming. As a result, residential/lifestyle farmers find CRP and WRP attractive, since these programmes require little time. Given their life-cycle position, many retired farmers have land available to put into conservation uses. The same forces may also be acting on low-sales operators, who average 57 years of age and may be scaling down their operations. If an off-farm job and advanced age are major determinants of land going into conservation uses, it may be relatively easy to get smaller farms to enrol land in the programmes. Getting larger farms to enrol more of their land might require higher payments, if the opportunity cost of idling their land is higher.

Land Tenure

Farm operators leased 38 percent of their total farmland in 2002, down from 40 percent in 1997 and 43 percent in 1992, according to the census of agriculture. This decline may reflect increasing rental costs as parcels of land become smaller. Parcels of farmland available to rent tend to become subdivided with time due to division among heirs. Smaller parcels increase transaction costs to operators assembling land to expand their operations. Still, rented land as a share of total farmland is higher than the 35-percent rate that prevailed in the 1950s and 1960s.

About 38 percent of all farms rented land in 2003, 32 percent as part owners and 6 percent as tenants. Land leasing has changed from a way for beginning farmers to enter agriculture to a way for established farmers to access additional land. Renting allows farms to expand without the debt and commitment of capital associated with ownership.

In fact, about 17 percent of very large family farms are tenants, a larger percentage than in any other group. Conventional wisdom holds that farmland owners have a long-term interest in their land and thus are more likely than renters to adopt conservation practices. Soule and others (1999 and 2000) found this to be true among corn farmers, at least in the adoption of conservation practices that provide only long-term benefits, such as grassed waterways and strip cropping.

The situation was different for conservation tillage, which can increase profits in the short run by maintaining or increasing yields while reducing machinery, fuel, and labour costs. Cash-renters are less likely than owner-operators to use conservation tillage, but share-renters appear to act like owner-operators in adopting conservation tillage. Share-renters may have an incentive to adopt conservation tillage, if the landlord bears some of the costs that may increase under conservation tillage, such as herbicide expenditures. Share-landlords are also more likely to be involved in management decisions than cash-landlords, which may make share-renters act more like owners.

DEFINITION OF A FARMING SYSTEM

To design appropriate or relevant ways of helping farmers, it is essential to understand the conditions under which farmers are operating. They have in fact, very complicated farming systems. The factors that have an influence on what the farming system will be. The operator of the farming system is the farmer or the farming family. To farmers, the way in which they earn their living and the economic, social, and cultural well-being of their households are linked closely and cannot be separated.

The members of the farming household have three basic types of inputs: land, labour, and capital. Management involves allocating these to three different activities or processes, that is, crops, livestock, and off-farm enterprises.

In making decisions on how to allocate their inputs in producing one or more products, farmers have to make some difficult decisions. These decisions will involve using their knowledge to come as close as possible to fulfilling the goal(s) for which they are striving. These goal(s) may vary from farmer to farmer. Many farmers are likely to want to maximize their incomes, once they have made sure they are producing enough food to feed their family and have met other societal obligations. The resulting combination of products (*i.e.*, crops, livestock, and off-farm enterprises) they are producing with their inputs results from the farming system they have adopted. Nevertheless, the extent to which the farming system fulfils the goal(s) they have chosen will depend on the managerial skills of the farming family and its ability to make good decisions in the allocation of inputs in a very uncertain production environment.

However, some parts of the environment that influence what the farming system will be are outside the control of the individual farming family thus causing uncertainty as far as the farmer is concerned. The 'total' environment in which farming households operate consists of two parts: the technical (*i.e.*, natural or physical) element and the human element.

- The technical element determines the types of, and physical potential of; livestock and crop enterprises. For example, in some areas, the low level of rainfall allows sorghum, but not maize, to be grown on rainfed land. The technical element includes physical and biological factors. These often are modified to some extent through technology developments — for example, increasing water availability through irrigation Improving soil quality by adding fertilizer, breeding for yield stability during drought, etc.
- The farming system that actually evolves, however, is only part of what is potentially possible. The human element is important in determining what the actual farming system will be. The human element consist of two types of factors: exogenous and endogenous.

Exogenous factors (*i.e.*, the social environment) are largely out of the control of the individual farming family. These factors will influence what the farming family can do and can be divided into three broad groups:

- Community structures, norms, and beliefs.
- External institutions, which include extension, credit, and input distribution systems on the input side and markets on the output side.
- Other influences, such as population density, location, and infrastructure.

Endogenous factors, on the other hand, are those the individual farming household controls to some degree. These include the types of inputs, that is, land, Iabour, and capita]. It is important to recognize that these resources and managerial ability vary among households and regions. The resources vary on the basis of quantity and quality, both of which influence the performance and the potential of the system. In addition, these inputs or resources may or may not be owned by the household. Access to one or more of these resources may be on another basis of use, which may limit or restrict the ease or intensity of use and thus, in turn, affect the goals and performance of the farm family. Nevertheless, it is the fanning family that decides on the farming system that will emerge.

However, this system will be influenced and sometimes constrained by the technical element and exogenous factors. The farming system is obviously complex, and the results can vary greatly because of differences in the 'total, environment. These facts help explain why some technology thought to be relevant often has not been adopted, or when it has, why the degree of adoption has varied widely. Not considering the human element in agricultural research has contributed to many socalled 'improved' technologies being irrelevant.

Status, Goals, and Appropriate Technology

Typical limited resource farm households in low income countries generally are characterized as follows:

"Farm households, with access to their means of livelihood in land, utilizing mainly labour in farm production, always located in a larger economic system, but fundamentally characterized by partial engagement in markets which tend to function with a high degree of imperfection."

Important points to note about this definition are that:

- Such households often are subordinate to some other external forces. That is, they do not have complete control over their own destiny and are in a process of transition. This is because of not being integrated fully in the market economy.
- They have access to land to pursue their livelihood. They cultivate the land largely with family labour, in conjunction with only small amounts of capital.
- The input or factor markets (*i.e.*, land, labour, and capital) work poorly. Concerning land, non-market rights of access or non-price forms of tenure are more likely to operate than a freehold market. The capital markets usually are developed poorly and variable production inputs are often not available. As a result:

"Credit and interest rates may be tied to other factor prices like land and labour within a dependent economic relationship. Thus factor markets may be locked together contractually rather than being independent."

Also, market information may be highly imperfect (*i.e.*, not available or readily accessible). Thus, it is not surprising that sharing and reciprocity often exist between households. In these non-market transactions, exchange of unlike goods and services take place, which are not valued by market prices. Sometimes, such relationships exist between households that are related to each other.

- Another reason why such households are integrated only partially into the market economy is because they consume a proportion of the product they produce. This enables them to have some ability to survive independently of the larger system, which may be important in explaining their economic behaviour.

  A number of microeconomic theories have been developed in an effort to explain the economic behaviour of farm households. These are based on various assumptions about household goals and the characteristics of the markets within which households make their decisions. Ellis [1988] notes that these theories are not mutually exclusive, sharing as they do certain key assumptions such as:
- The household is a single decision-making unit for economic analysis purposes. As a result, it maximizes a single utility (satisfaction) function that represents the joint welfare of its members.
- Profit maximization and utility maximization coincide where income is the only variable in the utility function. Profit maximization is always one of the components of utility maximization" when all the input and output markets are fully formed and competitive.

Ellis [1988] suggests that the theories predict different results because of different assumptions about the working of the factor and product markets, rather than because of differences in assumptions about household goal(s). In fact, an important characteristic distinguishing many of the theories results from different assumptions about labour markets and the allocation of household labour time, In support of these assertions, Ellis [1988] then discusses a number of these theories of economic behaviour. He calls these profit maximization, risk aversion, drudgery aversion, farm household theories, and share tenancy. In doing so, he recognizes the impossibility of separating short-run household decisions from the wider social relations of production.

Consequently, there is likely to be a certain universality in the aspirations of limited-resource farming households. The most important elements are income, effort avoidance, and risk avoidance. In economic terms, this means such households try to increase their utility or satisfaction, which increases with income but decreases with greater effort or higher levels of risk. This can be restated as maximizing income for any given level of effort or risk. Alternatively, it can mean reducing risk, or effort, for a given level of income.

Attempts to maximize utility or satisfaction take place within a set of constraints. As implied above, differences in the constraints, rather than the aspirations, lead to the most important differences in farming systems. Indicates some of the determinants of the farming system. It is easy to visualize constraints relating to the technical element causing differences in the type and productivity of farming systems, However, also important in the differentiation process are the factors relating to the human element. The quantity, quality, and ratios (*i.e.*, particularly between land and labour) of the endogenous factors (*i.e.*, factors of production — land, labour, and capital) play an important role in further differentiating farming systems. The extent to which they can be modified will depend on the degree of integration into the factor markets (*i.e.*? input markets). Also the products resulting will be influenced to some extent by community norms and beliefs and the degree of integration into the product market.

The above discussion has important implications for what is appropriate technological change. For example, limited resource households often are likely to be very little above survival level and, hence, will place a very high premium on improved stability of production.

The levels, qualities, and relative proportions of the factors of production are important in indicating the most appropriate route for improving the productivity and profit of the farming system — which is maximizing the return to the most limiting factor. For example, in areas of high land/labour ratios (*i.e.*, low population densities), labour saving types of technology/ systems are likely to be most appropriate.

In contrast, in areas of low land/resident ratios (*i.e.*, high population density) land saving technologies/systems will likely be more appropriate. However, increasing the return to the most limiting factor can have an indirect positive or negative impact on the use and productivity of other inputs. Thus, in view of the imperfect operation of the factor markets, care has to be taken in evaluating what would be the most relevant technologies.

For example, in a semi-arid area, initial superficial examination may indicate that because of the high population densities, strategies designed to increase the productivity of land would be most important. However, given the seasonal nature of agriculture in semi-arid areas, there will be periods of intense labour activity during the year. These will create labour bottleneck periods, followed by periods of underemployment. Thus, land-augmenting strategies using labour at critical labour bottleneck periods — when the opportunity cost of labour is high (*i.e.*, there are good alternative uses for labour that can yield an equally high return) — could be unattractive to the household. This is likely if adoption of such strategies involves the use of more labour at such labour bottleneck periods. It will also be likely if the productivity per unit of labour applied at that time is higher in alternative uses.

The results indicated the superiority of the improved maize technology over both improved or indigenous practices for sorghum and cotton in terms of both return per unit of land and labour. Note that for both the sorghum and cotton improved technologies, the return per unit area was higher than that from employing indigenous practices; but that the return per unit of labour used during the June/July period was the same or lower. Unlike the sorghum and cotton improved technologies, the improved maize technology was, from a profit viewpoint, suitable for farmers whether they were faced with a labour or land limitation.

Not surprisingly, since the 1970s, improved maize technologies have been adopted widely in Northern Nigeria, in a mini-Green Revolution. Although, initially the prime motivation for growing maize was as a cash crop, it eventually became a more important dietary item as farming households made trade-off decisions between growing the high yielding maize rather than the low yielding sorghum.

INTRA-HOUSEHOLD OR GENDER ISSUES

Many development projects have assumed that the male head-of-household provides the majority of agricultural labour and is the sole decision maker concerning agricultural activities. This approach fails to recognize the important contribution to women and children to agricultural production in many parts of the world.

Although data are very unreliable, the indications are that female family labour ranges from 2 to over 50 percent of farm labour.

Women's participation in production agriculture may range from zero in some Moslem countries to the situation in Africa south of the Sahara, which is a region of 'female farming systems' where women may manage and work more than one-third of the farms. Despite the great contribution made by women, development projects have been slow to recognize this contribution. As Jacobson points out:

"The reason is brutally simple: women perform the lion's share of work in subsistence economies, toiling longer hours and contributing more to family income than men do. Yet in a world where economic value is computed in monetary terms alone, women's work is not counted as economically productive when no money changes hands."

She goes on to suggest that this gender bias is compounded in many societies by discrimination based on class, cast, or race.

This situation can lead to complications because of differentiation in the types of labour available from household sources, Microeconomic theories that make simplistic assumptions about farming households add to the problem, For example, they assume a single decision-making unit and a single utility function representing the joint welfare of family members. However, increasingly this has been recognized as a problem, because intra-household relationships can have a major impact on economic behaviour, Some important points to note about intrahousehold relationships are as follows:

- Such relationships are of particular importance in situations where there is a high degree of congruence between the production and consumption units. They are also important where there is some imperfection in terms of integration with the factor and product markets.
- The economic situation of individuals within households is influenced by the social norms or behaviour of the society within which they reside, Therefore, specific economic roles are socially assigned to men, women, and children, Also, division of property, labour and income within households is socially, and not biologically, determined. As a result, males and females within households sometimes may have different income streams, with different sources and different destinations.
- Thus, it is wrong to assume that male and female labour can substitute for each other across the range of household activities. As a result, rigidities in time allocation may reduce the ability of households to respond to market signals. Therefore? such division of labour can impose constraints on seasonal labour inputs and ability to respond to price changes. As a result, it can have a major impact on the composition and level of farm output.
- Because of the above, increases in household income are unlikely to benefit all members equally. This often is exacerbated by the social subordination of women in many societies. As pointed out above, agricultural policies and projects often have ignored the role of women in agricultural production — not only as operators but also as decision makers. As a result, it is not surprising that technological innovations have tended to disadvantage women relative to men.

Therefore, an understanding of operational constraints facing limited resource households, and the relationships that exist between members of the households and between households is critically important in determining what are likely to be appropriate criteria for evaluating potential technologies. In connection with this, it is also important to understand the farming household members' goals and incentives, farming and non-farming activities, and resources available.

RECOMMENDATION DOMAINS

A recommendation is information that farmers can use to improve the productivity of their resources (*i.e.,* farming system). A good or successful recommendation, therefore, would consist of the practices that farmers would follow, given their current resources. If they had all the information actually available to the researchers. Obviously, the development of recommendations for farmers should be as efficient as possible. Because it is impossible to make a separate recommendation for each farmer practical compromises have to be made. As indicated earlier, this is done by stratifying the farmers into groups as homogeneous as possible. Farmers in these groups should have similar circumstances, resources, problems, and solutions to those problems. In this situation, therefore, it is usually assumed that the same recommendation will be suitable for all the farmers in the group. Such a group of farmers is a particular recommendation domain. Recommendation domains may be defined by agro-ecological (technical) and/or by socio-economic (human) circumstances. The definition of the recommendation domain depends on the particular recommendation.

Therefore, by dividing farmers into homogeneous groups or recommendation domains, one is recognizing:

- The heterogeneity that exists in the farming population.
- The likelihood that farmers, within a group, will have similar problems.
- The increased probability that farmers within a group, will be interested in the same potential solution to the identified problem(s).

Thus, the purpose of the groups is to highlight similarities within the recommendation domains and the differences between and among the recommendation domains. Similarities and differences are important only with respect to the objectives: increasing agricultural productivity by establishing experiments and generating recommendations. Because farm families differ from one another, it is important to distinguish the characteristics on which they vary, how those variables relate to the problems they have, and the usefulness of solutions to resolving those problems.In the descriptive/ diagnostic exercise, one of the purposes is to tentatively define the potential recommendation domains, although these may need to be adjusted in the light of later experience. An interesting short-cut that does not require a very precise definition of recommendation domains is the use of research-oriented farmer groups. Briefly, this involves offering to the farmers a number of technologies, from which they select those they are interested in testing. The characterization of the group selecting the option provides a quick indication of the possible recommendation domain for the proposed technology and reduces the chance of possible errors in the definition of the recommendation domains.

Inter-Generational or Sustainability Issues

Incorporating equity issues relating to the intergenerational impact of development is one of the challenges facing FSD. There is a great deal of concern about the impact of an ever increasing world population on the world's natural resources. This concern was evident at the Earth Summit held in June 1993 in Rio de Janeiro. Part of this concern was related to the loss of nearly 500 billion tons of topsoil through erosion in the 20 years between the previous meeting in Stockholm and the Rio de Janeiro meeting. According to Brown [1993], this and other environmental factors have contributed to a decrease in the rate with which food output is growing — an alarming situation given the increasing population to be fed. Most farmers, and hence development programmes, are concerned with local short-run production problems rather than the broad picture presented at the Earth Summit. This its logical, given the farmer's need to survive at the present, but it does not excuse the FSD team from considering the long-term, intergenerational and sustainability, impacts of their programmes, Although not the only factor to consider in the inter-generational and sustainability area, a major concern is soil conservation,'' Norman and Douglas [1994] identify a number of concerns when addressing soil conservation issues.

- Good land husbandry requires a mix of improved farming practices integrated with farm policies and programmes. Thus, it is essential to understand the current farming system in order to design effective conservation measures. This implies that a systems perspective be used.
- The goal should be to prevent rather than cure soil erosion. This means that the causes of soil erosion must be understood and treated, rather than concentrating on treating the symptoms.
- To be most effective, conservation programmes should be planned with the full knowledge and participation of the farmers. FSD team members must be able to communicate effectively with farmers to identify their ideas, perceptions, and priorities, so these can be taken into account in developing strategies for improvement.
- If farmers are going to implement changes, they must be convinced that they can obtain short-term benefits from change (*i.e.,* have an incentive to alter present practices). Thus, it is important to try and identify strategies that will have short-term benefits as well as provide long-run conservation improvements. When dual-purpose strategies cannot be designed, the long-term conservation activities should be linked with measures that provide short-term production/welfare benefits.
- For any conservation activities that require community action, it is important that the communities have responsibility and control and also reap the benefits of such action, in an equitable manner.
- Long term programmes are more effective than short-term programmes. Thus, it follows that the change will take place over a long period, necessitating adjustments and iterations in programmes.

Balancing Individual Farm and Community Objectives

Rapid changes are occurring in cropping system recommendations and practices in the Andes of Bolivia. Some of these changes are in conflict with objectives and beliefs held by the community. Individual villagers in this region occupy field sites for two or three years within communal lands and then shift to new sites. Individuals make their own decisions about cropping practices, and the community collectively decides when shifts are in order. The fallow period typically lasts from 3-15 years and provides communal grazing, rejuvenates soil fertility, aids with pest control in crop production, and provides fuel for the household. Furthermore, the long-term fallow system, aynuga, contributes symbolically to the well-being of the community and its sense of stewardship.

Adoption of 'improved, cropping practices by individual farmers on their fields could change the way in which they wish to occupy space in the long-term fallow system. Solutions to this restraint will require balancing objectives of individual farmers and community. These considerations are in addition to technical considerations in soil conservation and imply that the FSD team must possess skills other than just technical skills. This is implied in the example given in Box 4.2. Although this discussion has centred on soil conservation, the approach and factors to consider can he applied equally well to other programme areas where there are inter-generational or sustainability considerations-which will include most programme areas.

Participatory Research

During recent years, the farming systems approach has shifted dramatically from an emphasis on farmer based but researcher managed studies towards methods that stress farmer participation, or say, in the research process. The question asked is 'who is best positioned to lead an inquiry into a farming system'?'. Farmer-based, researcher-managed studies are necessary when:

"Farmers may not have adequate technical explanations of their problems nor know the range of opportunities for improving their conditions, but learning more about farmers helps the researchers produce better technologies and extension workers promote FSR results more effectively."

However, participatory techniques have proven very useful in allowing farmers to express themselves on a wide range of issues. In the extreme, the purposes of a participatory orientation could be simply to organize a voice for farmer concerns and to impact on political decisions. Another focused objective might be to establish nearly autonomous farmer groups that can be self-motivated and self-directed in problem-solving investigations, But, in practice, most participatory activities involve some balance, or a partnership, between researcher management and farmer say in the research dialogue.

The justification for shifting towards participatory techniques lies in the need to give farmers or groups of farmers the freedom to conceptualize and express an accurate model of how their farming system works, Participants can synthesize and exploit a comprehensive farm model in terms of systems structures, processes, and farm objectives certainly more efficiently, and often more effectively, than can FSD teams who study the system from the outside.

When embarking on greater farmer participation, FSD can choose to invite farmers to manage or comanage all or only some stages of the research effort, These stages might include: descriptions of the system, diagnoses of constraints, designs of new options, pre-screening or other assessments of options, selections of options for actual testing, reporting assessments of options to FSD, and even taking charge of disseminating useful information to other farmers. The role of FSD team members in participatory research usually is not reduced. In tact, it may require a greater awareness of issues (*i.e.*, farmers sensitivities, social customs, and technical understanding) than in more researcher managed work. In participatory research, FSD members should search out ways to:

- Facilitate the articulation of the participants' positions— often consensus of a group.
- Present hypotheses for discussion or testing.
- Link discussions with information from secondary sources that is not being taken into consideration by the participants.
- Communicate findings to outside agents.
- Link findings to the larger agricultural research and development context.
- Bring outside decision makers directly into the participatory process.

The correct balance between participatory methods and farmer-based, researcher-managed methods must be determined by the FSD team based on what is appropriate for each activity. The choice should be for the method or combination of methods that will give the greatest problem-solving efficiency. The choices, especially when a large degree of uncertainty exists about the issues and about the types of responses that can be anticipated, might be arrived at through an iterative process in which FSD teams routinely re-evaluate the techniques in their work plan.

Farmer participation is discussed or implicit under a number of the specific methodology sections of this manual. Obviously, a strong participatory role occurs in informal survey work and more particularly for PRA. A range of participation in trial management is discussed with emphasis on how this participation corresponds to the objectives of the trial. A participatory mode is greatest in FMFI trials. Farmer groups, listed under trial management methodology, are actually useful participatory forums to address essentially all stages of the research effort.

CONCEPTUALISING THE FARMING SYSTEMS DEVELOPMENT APPROACH

The material in Part I is designed to provide background information that will be useful in placing the more technical discussion in Parts II and III on the implementation of the FSD approach in context.

EVOLUTION OF THE FSD APPROACH

In the mid-1960's, there was little interaction between technical scientists (who were mostly on experiment stations) and social scientists (who tended to be concentrated in planning units).

The Green Revolution was beginning to have a great deal of success in Asia and Latin America, being based on good climate (*i.e.*, plenty of water) and soils; very homogeneous and favourable production environments; and the adoption of Improved varieties of wheat, maize, and rice that were very responsive to fertilizer.

Improved inputs also were readily available and there was an accessible market for the products. However, in most of Sub-Saharan Africa, and certain parts of Latin America and Asia, there has been no Green Revolution. This is because climatic conditions are often not as favourable (*i.e.*, too much or too little rainfall and limited amounts of irrigation), soils are generally poor, production environments are very heterogeneous and poor, and the input and output markets are poorly developed. Not surprisingly, there has been great difficulty in developing improved technologies that are attractive to farmers in such areas.

In Green Revolution areas, farmers were able to benefit from the improved technologies even if they did not do things quite right and the inputs they used were very divisible. However, in areas with less hospitable environments, farmers have to do things exactly right if they are to benefit and also they need lumpy inputs. In addition, yield increases are not so good — they tend to involve incremental rather than major changes in yields.

Thus, in the Green Revolution areas, because of the spectacular nature of the technology, experiment-station based technical scientists were very successful in their work.

However, the lack of success in using a similar approach in poorer agricultural areas (*i.e.*, with resource poor farmers), led to the evolution of the FSR approach, in which there is close cooperation between technical and social scientists. Work done with farmers in various countries in the 1960's and early 1970's revealed that these limited-resource farmers:

- Are rational (*i.e.*, sensible) in the methods they use. For example, in Africa, there was little support for station-based research on mixed cropping until the early 1970's. although earlier farm-based research had revealed the rationality of the practice.
- Are natural experimenters. Obviously, the methods farmers naturally use will be those that appeal to them and are informal in nature in the sense that they are not usually amenable to formal statistical analysis.
- Understand the environment in which they operate rather complex farming systems, consisting of crops, livestock, and off-farm enterprises. In fact, it could be asserted that such systems are often

more complex than the specialized farming systems found in many high income countries. Unlike the case with limited resource farmers in low income countries, many of the constraints in specialized agriculture in high income countries can be broken or avoided through seeking advice and taking advantage of and receiving external help.

- Consequently, considerable respect developed for limited-resource farmers. The FSR approach evolved because of an increased awareness on the part of researchers that such farmers:
- Had a right to be involved in the technology development process, because they stood to gain or lose most from adoption of the technology.
- Could productively contribute to the development of appropriate improved technologies.

Therefore, the fundamental principle of FSR was that farmers could help in identifying the appropriate path to agricultural development. It is now recognized that limited resource farmers can be involved productively in all stages of the FSR approach. Farmers' participation at all stages relates in one way or the other to the selection, design, testing, and adoption of appropriate technologies.

FSR rapidly became popular and was strongly supported by many donor agencies. By the mid 1980s, about 250 medium-and long-term projects worldwide were carrying out FSD type work. As one of the major donors, USAID between 1978 and 1988 had funded 76 bilateral, regional, and centrally funded projects containing either a farming systems orientation or clearly focusing on farming systems type work. Forty-five of these were in Africa.

Now, many of those projects are being institutionalized within national programmes with considerable domestic financing. Thus, the FSR approach evolved primarily as a result of a lack of success in developing relevant Improved technologies.

The strong technical focus that characterised the evolution still persists to this day, although increasingly many, including FSD practitioners, are advocating that the approach can be used constructively in addressing not only technological solutions but also those relating to policy/support systems, a topic that will be considered further later in the manual.

In the mid-1980s, the Farm Management and Production Economics Service of FAO became actively involved in the farming systems movement and developed the FSD approach. FSD is based on the farm-household focus of FSR and emphasizes the central role the farmer plays in farming systems development. In addition, this approach incorporates an increased emphasis on the dissemination of improved farming systems based, in the process, on active involvement of those responsible for the policy/support systems. In the rest of the manual the term FSD will be used instead of FSR in order to avoid any possible confusion.

COMPARING STATION-BASED RESEARCH AND FSD

The literature frequently compares and contrasts station-based research and FSD. The tone in such comparisons often implies that they are substitutes for each other. However, both station based research and farm-level FSD are needed. This is because they focus on different things that are complementary to each other.

A major difference between them is the following:

- On the experiment station, applied research is usually undertaken, in which new technologies are created.
- FSD, on the other hand, concentrates mainly on adaptive research, which involves helping to adjust technologies to specific environmental conditions. FSD also helps feed back information about future priorities for applied research to experiment stations.

Some of the major differences between station-based research and FSD and helps indicate why both are necessary, particularly in areas where the Green Revolution has not occurred. In such areas, greater attention will need to be paid to adaptive research, if relevant improved technologies are to be developed and adopted by farmers.

Obviously, strong linkages between these various contributors are of crucial importance. The FSD approach, involving working directly with the farmers, is a more 'bottom-up' or 'micro to macro, orientation, compared with the more 'topdown, or 'macro to micro' orientation of research work or planning exercises that start at the experiment station or the upper levels of planning ministries.

Part 'B' indicates that FSD can help in strengthening 'bottom up' linkages amongst the various groups. In doing so, it helps in the process of improving productivity but does not result in a product by itself. (*i.e.*, facilitates a process rather than producing a product).

Relevant improved practices/technologies and policy/support systems are needed to bring about increased agricultural productivity. Decisions on which strategy is to be emphasized to increase agricultural productivity will depend on circumstances.

For example, if no technology is available to ensure an economic return to fertilizer, there will be little value in concentrating on developing an input distribution system that makes fertilizer available to farmers. Therefore' the complementary nature of the relationship between developing and disseminating relevant improved technologies and policy/support systems needs to be constantly borne in mind, if the agricultural development process is to proceed efficiently.

It is apparent that the FSD approach has a broader focus than simply addressing technological issues— the primary focus of early FSR activities. The FSD approach, can help also in facilitating linkages not only between farmers and station-based researchers but also with other 'actors,, including those responsible for designing and implementing the policy/support system. In fact, most work in the farming systems arena to date has concentrated on the technology thrust or farming systems activities relating to FSD. This manual concentrates most attention on the research or technology thrust of FSD although linkages with other 'actors' besides farmers and station-based researchers are discussed to some extent.

GENERAL APPROACHES TO THE FARMING SYSTEMS APPROACH TO DEVELOPMENT

APPROACHES WITHIN FARMING SYSTEMS RESEARCH

Because FSD has its roots in farming systems research, the following comments on approaches within FSR are applicable to the technology research activities of FSD. The term FSD often is used loosely. An activity is considered to be a legitimate part of the technology research function of FSD if the following situations hold:

- The whole farm is viewed as a system.
- Research is conducted with a recognition and emphasis on the choice of priorities that reflect the whole farm.
- Research on a farm sub-system is legitimate FSD, provided the connections with other sub-systems are recognized and taken into account.
- Evaluation of research results explicitly takes into account linkages between sub-systems.
- As long as the concept of the whole farm and its environment is preserved, not all the factors determining the farming system need to be considered as variables — some may be treated as parameters or constants.

A distinctive feature of FSR, or rather FSD, is that a systems perspective is constantly borne in mind. Within this framework, one can look at either a small or a large number of variables in the system. In general, because of the complexities of simultaneously handling a large number of variables, most FSR programmes have tended to limit the number of variables they study and to regard the other factors that influence the farming system as parameters or constants.

Three general approaches to FSR or FSD with a technology generation focus, have been defined relating to the number of variables being investigated. These approaches can be described as FSD 'with a pre-determined focus', 'in the small', and 'in the large', In a sense, these types represent an evolution in FSD as techniques have been developed to handle progressively more complex situations— defined as involving a higher ratio of variables to parameters, The progression, which is illustrated, where it has been expanded to four phases, can be articulated as follows:

- In its early days, FSR (*i.e.*, FSD) type activities concentrated on how farm yields of particular crops could be raised. This form of on-farm research stood in sharp contrast to earlier multi-locational on-farm trials, in which only the effects of the physical (*i.e.*, natural or technical) environment were tested, Even though this FSD approach did involve the inclusion of socio-economic elements and, hence, had a farming systems perspective, it was done with a 'predetermined focus' on the productivity of a particular commodity. Thus, this approach involves looking at one facet of an enterprise or one specific enterprise and identifying improvements within that focus that are compatible with the whole farming system. With the creation of distinct FSD teams — in contrast to an on-farm testing component located within each station based commodity team — FSD activities generally have evolved beyond this level. However, it is important to recognise that this type of activity still is a legitimate part of any FSD in the sense that it can provide a very useful input into the activities of commodity research teams.
- In the pursuit of greater farmer participation, research shifted to addressing farmers' articulated problems directly. Still, farmers' problems were very much con fined to crop production, although links with other components of the whole farm system were studied. An alternative name for this 'farming systems in the small' focus could be 'farming systems with a whole farm problem focus', Many FSD efforts currently tend to be at this point, However, currently, on-farm research projects that seek to improve whole farm performance through manipulation of the linkages between all enterprises are :
    - Looking further ahead and to enable addressing issues of sustainable agriculture effectively, concerns for maximizing the economic and biological performance of enterprises will be modified by concerns relating to ensuring sustainability, including the rehabilitation and regeneration of natural resource systems. This natural resource system dimension to sustainable agriculture can be thought of as a type of 'farming systems in the large, or 'farming systems with a natural resource systems focus',
    - Similarly, attention to ecological sustainability will need expanding to include the many off-farm and non-farm activities that make up sustainable livelihoods. This also can be thought of as a type of 'farming systems in the large, or 'farming systems with a livelihood systems focus'.

These notions, particularly the last two, have been largely unexplored until recently, when techniques have evolved to handle complex situations in which the ratio of variables to parameters is high. Formal modelling techniques by themselves generally have limited value in addressing such situations because of the complexity of the relationships and the degree of understanding that is required initially to develop a realistic model. However, the major breakthrough that has occurred is the ability to pursue a less formal or more informal modelling approach through application of farmer participatory type techniques that, in essence, involve using farmers' minds as computers!

Introduction to FSD

Details concerning the characteristics and methodology of the FSD approach, as developed by the Farm Management and Production Economics Service of FAO, are available elsewhere. Thus, the following discussion represents only a brief summary. The primary objective of FSD is to improve the well-being of individual farming families by increasing the overall productivity of the farming system in the context of both the private and societal goals, given the constraints and potentials imposed by the factors that determine the existing farming system. It is based on the development principles of improving productivity, increasing profitability, ensuring sustainability, and guaranteeing an equitable distribution of the results of production,

The farm household is the principal system and focus of FSD and consists of three basic subsystems, which are closely interlinked and interactive: the household, the farm, and off-farm activities. The two major categories of activities or thrusts of FSD, both of which involve intensive interaction with farmers? are:

- Farming Systems Analysis. This involves studying, together with the farmers the natural (*i.e.*, technical) and socio-economic (*i.e.*, human) environments in which farm households operate. The aim is to identify the constraints limiting farm productivity and production and hindering improvement in the welfare of the farm households themselves. Potential solutions to these problems are identified, and the results of this analysis, formulated as suggestions for further action then are passed on to the relevant 'actors'. These could include researchers, extension and support service staff, and/or policy makers.
- Farming Systems Planning, Monitoring, and Evaluation. These involve testing, monitoring, and evaluating improvements on-farm, with the direct involvement of farmers. Examples of such activities include those regarding proposed technological improvements, proposed revisions in farm plans, and improvements in support services and farm-level impact of proposed policy changes. Those improvements thought to be potentially useful then are disseminated to other farmers via the extension service. Prior to this stage, it may be necessary to negotiate adjustments in the policy/support services that will facilitate the dissemination and adoption of the improvement. After dissemination, it is of primary importance to monitor and

evaluate the adoption rate of the proposed improvements by the farming community. This provides an indication of the actual value of the improvement being disseminated.

- Many individuals in the last 15 to 20 years have written extensively to clarity the concepts of the farming systems, particularly with respect to the development of relevant improved technologies. There are many ways of presenting a diagram involving the application of farming systems methodology to research (*i.e.*, technology development). One such way of emphasizing the technical research aspects of FSD. There are four fundamental stages in the FSD approach.
- The Descriptive or Diagnostic Stage, in which the actual farming system is examined in the context of the 'total, environment — to identify constraints farmers face and to determine the potential flexibility in the farming system in terms of timing, unused resources, etc. An effort also is made to understand the goals and motivation of farmers that may affect their efforts to improve the farming system, The following sequence of activities is usually undertaken:
    * After deciding on the location of the work, start-up activities include reviewing secondary sources of information, making the necessary contacts, assembling the professional team, and making logistical arrangements,
    * Farming families are classified tentatively into homogeneous groups. The farming families within each group or domain usually practice the same farming system(s), face the same constraints, and have the same potential solution(s) to their problems. This tentative classification can be modified as additional information becomes available.
    * An informal exploratory diagnosis or survey is undertaken to obtain a qualitative understanding of the determinants of the existing farming systems (*i.e.*, both biophysical and socioeconomic). At the same time, efforts are made with farmers to ascertain the constraints, flexibility, and potential opportunities in the farming systems they are using currently.
    * Sometimes a verification survey involving a structured formal survey, suitable for statistical analysis, is undertaken to quantitatively confirm insights obtained in the exploratory diagnosis. It also can be used to provide a database for farm and development planning, policy analysis, monitoring, and evaluation.
- The Design Stage, in which a range of strategies is identified that are thought to be relevant in dealing with the constraints determined in the descriptive or diagnostic stage.

The usual sequence of activities involves the following:

* Together with the farmers, the constraints are ranked according to their severity, and potential solutions are identified after determining what flexibility exists in the farming systems currently practiced. Information for designing such strategies (*i.e.*, particularly relating to technological issues) comes from experiment station work, researcher managed and researcher implemented (RMRI) type work on farmers' fields, and from other farmers.
* An evaluation is made of the proposed solutions before putting them into practice. These solutions can be technological or institutional in nature. A number of analytical techniques can be used to evaluate the potential technical feasibility? economic viability, social acceptability, and ecological sustainability of the proposed solutions before they are put into practice. One or more of the proposed solutions are selected for actual evaluation on-farm.

- The Testing and Implementation Stage, in which one or more promising strategies, arising from the design stage, are examined and evaluated under term conditions to determine their suitability for producing desirable and acceptable changes in the existing farming system.

Activities at this stage involve the following:

* On-farm testing or evaluation with farmers determining how well the potential improvements fit into the system, whether or not they are acceptable to farming households, and what modification may be needed to make them acceptable. In terms of technological issues, this stage conventionally has consisted of two steps:
  — Researcher managed but farmer implemented (RMFI) tests to establish whether previously determined technical relationships are altered by farmers' management of non-treatment variables.
  — Farmer managed and implemented (FMFI) tests, when the team is confident that technical relationships will hold but needs to evaluate the proposed technologies under local socio-economic circumstances. In addition to the on-farm evaluation of the proposed technologies, evaluations also can be made of the proposed farm plans and of proposed changes in support systems and/or policies.
* Positive results of such evaluations provide justification for the FSD team to advocate further action. The tested technologies can be disseminated through the extension service to other similar farming households, The same applies to the farm plans that have been tested successfully. Favourable test results for proposed

changes in support services and/or policies provide valuable farm-level information on necessary programme or policy adjustments.

- The Dissemination and Impact Evaluation Stage, in which the strategies that were identified and screened during the design and testing stages are extended to farmers. In terms of activities at this stage, impact/adoption studies can be very important. These potentially can be very useful, not only in giving some idea of the impact of, for example, agricultural research, but also in giving some idea of future priorities for agricultural researchers and indicating what adjustments are required in the policy/support systems to ensure better rates of adoption. Thus, through monitoring and evaluating the impact and rate of adoption of the changes that have been implemented and proposed/tested earlier, it is potentially possible to provide some indication of further desirable activities by both researchers and planners.

There are often no clear boundaries between the various stages, Design activity, for example, may begin before the descriptive and diagnostic stages and may continue into the testing stage, as promising alternatives emerge from RMFI trials.

Similarly, testing by farmers may mark the beginning of dissemination activities. Also, going through all stages may not always be necessary. FSD team confidence in transferability during design/planning activities can sometimes mean going straight to FMFI work or even to recommendation/ dissemination activities.

Thus, the process of FSD is recognized as being dynamic and iterative, with linkages in both directions between farmers, researchers, extension staff and policy/support service staff, The iterative characteristic can improve the efficiency of the development process by providing a means of identifying and fine tuning improved technologies for a specific location-that is, climatic situation, soil type, and/or farmer resource base.

Fsd with A 'Pre-determined Focus' on Specific Technology

Researchers in station-based programmes often want FSD workers to evaluate specific technologies they have developed to see how they fit in with constraints of the farming system currently used by farmers. This type of work is referred to as FSD with a 'pre-determined focus' or 'predetermined technology focus,' In Botswana, FSD teams and participating farmers have assisted station programmes in testing cowpea varieties, evaluating the desirability of hilling groundnuts, ascertaining whether the Dutch hoe can help with the weeding operation, and so forth.

This type of testing in FSD is not analogous to typical technical transfer. Rather than testing fully finished products, FSD assists technology generating programmes, usually based on research stations. Often, the types of technology tested in this manner have not been identified as having high leverage power in the system. Leverage power refers to the ability of a technology change to have a major impact on the performance of the farming system. However, these technologies do fit possible needs and asking farmers to participate in their development is worthwhile, It is desirable to test technologies as early as possible in their development, This is to facilitate making modifications and to minimize waste of research resources. It also may accelerate the dissemination of suitable technology, Such technology, even if not fully appropriate, may sometimes stimulate thinking among farmers if they are not accustomed to these types of interventions. Feedback in these instances can contribute to new innovations in technology generation.

FSD with a 'Pre-determined Focus' on a Subject

A great deal of FSD work, particularly in its earlier years, was not focused on testing specific technology as much as a pre-determined focus on a subject matter area. FSD within the CGIAR institutions such as IRRI, CIMMYT, and ICRISAT have focused on issues related to production of their respective commodity crops. ILCA focused on issues in livestock production, ICRAF in agroforestry' etc. Some FSD teams within the national agricultural research systems (NARS) also worked under limited mandates to deal with particular enterprises; commodities; or other subject areas. The judgements of farmers, the end-users of products developed by these organizations, were foremost in these FSD activities. However, the parent institution with its technologygeneration activities would be a primary client in these instances.

FSD with a 'Whole-farm Problem Perspective, or 'FSD in the Small'

An FSD interdisciplinary team conducted a diagnostic survey and determined that labour for satisfactory weed control in field crops is a major constraint to farm production and farm family well-being. The survey also showed that due to population pressures, farmers were no longer able to shift field sites as in the past and that deteriorating soil fertility also was becoming a major issue. However, during the course of farmers participating in testing possible solutions to these problems, discussions between the FSD team and farmers identified the potential for intensified poultry production as an additional means of improving farm income. This resulted, in addition, in collaborative testing in the area of intensive poultry production and, as a result, exemplifies FSD with a 'whole-farm problem perspective.'

The major breakthrough that has occurred is through supplementing the rapid rural appraisal (RRA) techniques, which have been available for some time with the participatory rural appraisal (PRA) techniques, which have developed rapidly in recent years. To date, the first two approaches listed above have been used. Although FSD 'in the small' arrives at a focus within the system in the course of diagnosis, FSD 'with. pre determined focus' moves into the system to research an enterprise or one facet of an enterprise looking for improvements within that focus that are compatible with the whole farming system.

FSD WITH A 'NATURAL-RESOURCE MANAGEMENT FOCUS

A non-governmental organization (NGO) has adopted, as one of its aims, the conservation of the natural resource base within areas of agricultural production.

This NGO sponsors FSD activities in a village in which participating farmers look at current and prospective problems in natural resource conservation and management. They evaluate potential solutions to these problems at the same time issues in agricultural production and family income are considered.

The FSD team from the NGO and farmer participants discovered a number of incongruencies between issues of conservation and present day agricultural production. They included the following:

- Natural-resource conservation analyses focus on the landscape, which in some instances does not even correspond to a village or to any other political entity. On the other hand, agricultural production analyses tend to focus on individual farms, units within farms, or smaller interacting networks of farms.
- Conservation invariably takes a long-term perspective of problems and solutions, whereas production analyses consider short-term returns and immediate survival issues.
- Natural resource conservation often involves community or ultra-community efforts and goals, which might conflict in terms of labour or other requirements, with individual farm production activities.

Integrating these apparently opposing sets of objectives for natural resource conservation and production poses a major challenge for FSD. Participating farmers recognize many of the conservation issues and trends and some express concern for the future.

In some cases, traditional techniques for conservation such as gully amelioration have been identified and are being promoted, but other issues remain unresolved. Some of the constraints encountered today or anticipated in the future were unknown to farmers in the past. Solutions therefore, are not known, and innovative and problem-solving thinking is required. However, given farmers' intimate knowledge of their current and past production environment, their input in designing appropriate strategies for the future based on an analysis of what has happened in the past is indispensable. This is clearly a fertile area for active collaboration between on-farm and on-station work: a give and take between farmers, scientists, and policy makers.

SOME CHARACTERISTICS OF FSD

Some of the major characteristics of FSD are as follows:

- Farmer Centre Stage. The farmer, as the consumer of the improved technologies, is in the centre of the stage. This provides an opportunity for researchers to learn from the farmer, enables him or her to have an input into the research process, and ensures that criteria relevant to him or her are used in evaluating proposed technologies. For the farming family, evaluation criteria (*i.e.*, for the adoption of the improved technologies) can be divided into the following groups:
  * Necessary conditions determine whether the farming family would be able to adopt the improved practices. Such conditions include technical feasibility, social acceptability, and compatibility with external institutions — that is, support systems.
  * Sufficient conditions determine whether the farmer would be willing to adopt the improved practices. Obviously, the necessary conditions will be influential in determining this willingness. Sufficient conditions include the compatibility of the improved practices with the goal(s) of the farming family; the resources they have access to; and the farming system they now practice.
- Work with Representative Farmers. Although the input of farmers is critically important in FSD it is impossible for FSD teams to work with all farmers. Therefore, a few are selected that are thought to be representative of all farmers. Because there are many types of farmers— with differences in the products they produce, the resources they possess, and the problems they face — it is necessary to put farmers with similar characteristics in the same group (*i.e.*, stratum or recommendation domain). A small sample of farmers in each of these groups then is selected to work with the FSD team. If the grouping and selection of farmers are carried out correctly, then the results obtained by these representative farmers should be achievable by other farmers with similar characteristics, when the technologies are extended to other farmers by extension staff.
- Involves an Interdisciplinary Approach. As indicated earlier, farmers have complex farming systems. As a result, changes in one part of the farming system may have a good or bad impact on another part of the farming system. For example, in some areas, it is not realistic, or indeed desirable, to try and increase crop productivity without considering livestock at all,' Also, attaining a successful solution to a technical problem that has been identified will depend on whether the farming family has the labour and money to adopt (*i.e.*, an economic issue) and any sociological reasons preventing them from adopting. Therefore, to address the wide range of farming systems' issues, FSD teams generally consist of representatives of a number of

disciplines — usually agronomists, animal scientists, and agricultural economists and sometimes sociologists. As a result, an interdisciplinary approach, that is a number of disciplines working together on the same problem, has to be used to solve the problems of farmers.

- Dynamic and Iterative Approach. Because of the dynamic nature of agriculture, research is a never-ending process. Farmers always face problems to varying degrees in their farming operation, and many of these problems can be solved through research. Sometimes the solutions suggested as a result of research don't work and need to be modified. The dotted lines indicate that such failures require moving back to an earlier stage in the FSD process and repeating one or more of the steps in the process. This makes FSD an iterative as well as a dynamic approach.
- Complementary to Station-based Research. In the context of technology choice and development, the role of FSD (*i.e.*, utilising a systems approach) is seen to be complementary to technical component research (*i.e.*, which primarily employs a reductionist approach), most of which is undertaken on experiment stations and is usually commodity-based, With respect to such research, FSD has three roles:
  * To look at recognized farming systems and the stock of materials and techniques accumulated from station-based research, so as to be able to choose technical solutions to problems that have been identified, On-farm experimentation then adapts chosen solutions to the local situation. This is a mobilizing and adaptive role, shaping the product for an identified market.
  * To pass back unsolved technical problems, important to the system, to the appropriate commodity research team on the experiment station, This role is one of identifying and helping to prioritize the agenda for technical research.
  * To link with farmer clients and extension staff in local farm situations, drawing both farmers and extension workers and other relevant 'actors' into the technology-generation process.

Two examples indicating the perils of not exploiting the complementarily between on farm and station-based research are given in Boxes 3.6 and 3,7. Similar linkages could be established with planning and development agencies concerning proposed policy/support programme changes. Unfortunately, these are not yet strongly developed in most countries.

FSD with A 'Livelihood-systems Focus

There are, to date, relatively few good examples of applying FSD with a 'livelihood systems focus.' One organization that in recent years has expended considerable effort in integrating 'off farm' with 'on-farm' activities in a sustainable manner has been ICLARM which has concentrated on integrating aquaculture with crops and livestock.

Certainly integration of 'on-farm' and 'non-farm' activities deserve much greater emphasis than it has to date, especially as increasing pressures develop on the natural resource base. 'Non-farm' activities car provide an outlet for surplus labour and contribute to the attainment of sustainable livelihoods. In many areas, particularly where agricultural activities are seasonal in nature, farmers have always recognized the complementarily between farm and non-farm activities, through emphasizing the latter particularly during slack periods in the agricultural cycle, Many of these jobs are in the non-formal sector and involve micro-enterprises, also an area that needs much greater emphasis than it has to date.

A Systems or Reductionist Approach?

Is this the only choice? It might be popular to think in terms of the one or the other; systems, approaches or reductionism, each with its characteristics and methodology. Should one 'divide and conquer' or accept that 'the whole is greater than the sum of its parts"? A distinction is made between science and art but also it is reflected in divergent schools of thinking within some disciplines. However, in FSD, this rigid dichotomy of perspective is neither necessary or feasible.

FSD recognizes that farming systems are highly complex and not fully reducible. They are too complex for human intuition to deal with all the factors involved but, at the same time, consist of such a fuzzy web of interactions that currently available analytical tools are not powerful enough to describe them.

Over the years, science in agriculture has attempted to model the factors and relationships within a farm operating in an agro-ecological situation. The purpose of this modelling invariably has been to engineer optimal solutions or strategies. An important limitation to this approach is that optimization has been for the research model and not for the farm and the farming family itself. A larger purposeful framework comprising a farm value system and related decision making patterns cannot be incorporated so readily into this research model.

FSD seeks to synthesize human intuition and judgement with analytical science. To accomplish this, FSD uses farmer perspectives on issues, problems, etc. Even when these are not completely exact, they help create holistic or logical frameworks within which technical alternatives can be screened or engineered. FSD also seeks to render the holistic framework as explicit as possible. Farmers and FSD staff seek to differentiate uncertainty in the system that is due to randomness (*i.e.*, rainfall probabilities) from that due to lack of experience (*i.e.*, ignorance about new technology) from that due to vagueness (*i.e.*, typically, decision making patterns). By integrating relevant scientific information, hands-on experiences, and intuitive managerial skills, FSD can be effective in evaluating scenarios, even those involving future changes in the farming environment.

Within this framework, numerous scientific information needs will be recognized. Within this framework, the need is to obtain rather unequivocal answers to the specific and well defined questions. Reductionist's techniques are better suited for this purpose, especially if the question pertains to objects in the system rather than human behaviour, which involves human intuition skills. The key to successfully designing this type of focused research is the incorporation of different perspectives through interdisciplinary teamwork and participation of farmer managers. Thus, 'rigour versus relevancy' need not be a dilemma in FSD, On the contrary, rigour and relevancy are both possible and essential, if solutions are to be found in time for the many pressing problems that confront farming systems in a rapidly changing world.

The systems analysis employed in FSD should do the following:

- Pay particular attention to formulating problems within the holistic framework.
- Review each result within the general systems framework and never exclude an alternative without having done this. Sometimes, the best alternative in a systems framework is tar from optimal when evaluated separately.
- Set forth hypotheses early, so that all 'actors' with their perspectives can participate in reformulating these questions. After all, it is these questions more than 'responses, that set systems approaches apart.
- Avoid overemphasizing mathematical representation at the earliest stages.
- Treat even uncertainty as explicitly as possible.
- Sub-optimize with care.

Thus, there is a complementarity between the reductionist and systems approaches. This is why it was argued earlier that the FSD approach is complementary to station-based research, which usually employs a reductionist approach. The holistic approach in FSD is needed to ensure that experiment stations develop relevant technologies and needs to exploit the power of the reductionist approach, which can be a very efficient way of developing new technologies, once the priorities have been ascertained correctly. The results of this potentially relevant technological work can then be given to the FSD team for on-farm evaluation.

Technology that Works Under Experimental Conditions

In Botswana, precipitation is low and variable, creating erratic conditions for crop production. Under these conditions, farmers minimize risk by adopting an extensive low-input cropping system that is moderately productive under favourable conditions but minimizes losses when rainfall fails. Under research-station management, an intensive system for maize increased average yields and dramatically improved yield stability. The intensive system consisted of controlled pathways for planting with minimal tillage of the soil and water concentration onto the plant zone. It is the water concentration that stabilizes maize yields. Weed control was as used needed but weed pressures were minimal when crops were established early.

This technology failed when it was tested under farmer management. It failed because critical resources are not available on the farm at the right time to meet the schedule requirements of the new technology.

The intensive system requires that soil preparation be finished before the rainy season begins and that planting be carried out on the first rain(s) of the year. Meanwhile, draught animals to do this work are kept near watering points at distant cattle posts until rains have filled shallow wells and ponds near the herds. Farmers and their draught animals cannot plant early in the season, and when they do not, weed growth overruns the controlled pathways. The Ministry of Agriculture made a fitting decision to conduct farmer-managed tests before recommending this system to farmers based on station results alone.

Appropriate Technology' will not Always be Relevant on-farm

Appropriate technology is characterized as low cost, technologically simple, labour intensive, making use of local materials and existing methods of production, and responding to identified constraints in the farming system. FSD generally seeks technologically appropriate solutions to problems but these should not be promoted automatically without on-farm assessments. The Bolivian altiplano is a hostile environment for agricultural production. The zone is dry and cold with nightly frosts occurring at any time of year. In earlier periods, farmers of this area traded products of their system for various food products of the lowlands. In recent times, their access to other markets has partially been broken off. Land degradation and pressures on use of the land have further reduced the ability of farmers to maintain their traditional crop and livestock production strategies. Protected horticultural systems were identified by several NGOs as appropriate ways to solve problems in this system. For the most part, these systems were promoted directly without an objective assessment including testing under farmers' conditions. Initial glasshouses were soon found to have difficulties with high insolation costs and were replaced by adobe structures covered with plastic. But many other problems were not addressed or proved unsolvable: guaranteed water supplies, local and distant markets for excess production, nutritional benefits of products, and so forth.

An important finding of a study of this large and costly effort is that promoters of development have not taken experimentation seriously enough. The value of farmer and expert involvement in design, testing, and dissemination is that irrelevant or incorrect designs and management systems can be screened out or corrected.

Challenges Facing FSD

A number of challenges face FSD if it is to continue playing a significant role in facilitating the process of agricultural development. Some of these are:

- Better Incorporation of Farmers, Incorporating farmers into the research process was one of the most important principles underlying the evolution of the farming systems approach. The basic justification for this was that farmers could improve the efficiency of the research process. At a minimum, it was argued they could prevent the use of research resources on types of technologies that they would be unlikely to adopt, while even more Importantly, it was anticipated they could, in their own right, contribute positively to the research process. Unfortunately, farming systems workers often have not sufficiently recognized this positive and interactive contribution of farmers. Consequently, this spawned the farmer participatory research (FPR) movement. The problem of not utilizing the farmer sufficiently in FSD activities is not due to inherent deficiencies in FSD itself but rather with the way in which it has been applied, Techniques of FPR or PRA need to be incorporated into FSD.
- Continued Evolution of FSD. FSD is relatively new. Therefore, the methodology is still evolving and, as a result, universally accepted 'standard texts' on the 'nuts and bolts, of how to do it are still to emerge. Related to the methodology issue is that time and cost-efficient (*i.e.*, money and people) methods for undertaking FSD still need further development. This is important because of the limited resources available for undertaking research, both on-station and on-farm.
- Greater Incorporation of the Policy/Support System Perspective. As has been stressed already' the farming systems approaches implemented to date have focused mainly on the technology dimension (*i.e.*, FSR). However, a basic principle of the FSD approach is that the farming systems perspective is critically important in formulating and adapting policy/support systems in ways that will facilitate and accelerate the agricultural development process, Once again, discussions relating to the policy/support system are scattered throughout the manual, and it is highlighted as an essential component of FSD. Failure to incorporate this dimension in FSD activities is likely to have an impact analogous to playing soccer on one leg, Nevertheless, because of the lack of much proven experience and documented material on the farming system perspective with respect to the policy/support system, this manual does not deal with

this dimension to the extent that would be desirable. Hopefully, this lack of experience will be rectified in the fairly near future. An example of the potential value of using a policy/support perspective in FSD is given in Box 3.8.

- Incorporating Equity Issues — Intra and Inter-Generational. FSD tries to help the farmer with the problems he or she has identified. Of course, the reason for this is the necessity to introduce an intervention in which they are interested, It its likely that these 'felt, problems of farmers are likely to have a short-run focus (*i.e.,* particularly when the farmer is operating very close to the survival level). Also, it is possible that helping to solve the individual farmers' problems creates others for the society as a whole. Although the example given in Box 3.9 pertains to inter-household relationships, equity considerations within a particular generation also apply to what is happening within farming families. It is wrong to assume that all farming households or families operate in such a way that distribution of effort and benefits is equitable. For a positive example of the benefits of understanding intra household relationships. When designing a technology to help farmers increase their productivity, consideration must be given to the possible long-term effects of that intervention. Therefore, if farming system (FS) workers are not careful, their work can result in creating two types of inequalities, that is, helping some farming households— or even certain individuals within those households — at the expense of others and/or reducing the quantity and quality of land that can be productively farmed by future generations. Avoiding the development of such inequities constitutes a major challenge to FSD teams who need techniques to be able to screen out proposed technologies and policy/support systems that would encourage or exacerbate such trends. Once again, issues relating to equity matters, although generally recognized as important, still need attention in farming systems type activities. This will require continued evolution of FSD towards the FS 'in the large' approach.
- Assessing Agricultural Research Impact. Related to the research resource issue is the importance of devoting some effort to assessing the impact of the research process-something that often has been done inadequately. More attention needs to be paid to adoption/diffusion studies. Such studies, of course, are the best measure of the impact of the agricultural development process, Therefore, the credit for favourable results from such studies cannot be allocated to FSD activities alone. However, FSD teams, because of their on-farm location and pivotal linkages with other actors, are often in the best position to take a leadership role in their execution. Also, as suggested earlier, such studies can and should be used for other purposes, such

as feeding back priorities for further research and providing evidence for adjustments in the policy/support system to encourage greater adoption,

- Improving Credibility of FSD. Establishing credibility for FSD-related activities is a major challenge and is necessary to ensure that some of the limited research resources always will be allocated to them, As argued earlier, FSD staff constitute only one set of the 'actors' in the agricultural development process. Additionally, FSD helps facilitate a process and does not result in a product by itself. Thus, FSD cannot claim sole credit for any technologies developed for, or adopted by, farmers. However, it achieves credibility through its linkages and cooperative efforts with other 'actors' in the agricultural development process. These 'intermediate products' need to be documented and publicized to a greater extent than often has been the case.

FSD in the Context of other Agricultural Institutions

Work relating to FSD might be incorporated as a working tool for several different types of institutions. These could include such diverse institutions as agricultural research, extension, community development (particularly in regards to NGOs), or even ministerial or national units interested in policy design and assessment.

FSD thus can he used to focus on family income. technology design, dissemination, and adoption: or designing solutions to the problems associated with a deteriorating natural resource base in agricultural areas. At another level of viewing the institutional setting of FSD, it is apparent that not only some disciplines but also various technical and sociological approaches and methods, are evolving to better deal with systems analysis.

Take, for examples, the use of Geographic Information Systems (GIS) as a tool integrating large amounts of data into a single framework useful for technical as well as sociological and economic analyses. On the other hand, community development and policy reform programmes use increasingly sophisticated techniques (*i.e.*, tools or methods) that contribute to decentralized planning, grass-roots participation, and empowerment in assessing problems and designing solutions.

The emphasis in many areas is on developing integrated approaches and system type analytical methods. With the possibility of increasing convergence occurring in some of the methods and, in some cases, objectives, the unique contributions by FSD will be brought into question by some.

And, in fact, the institutional role or mandate of FSD will vary. FSD in one setting may focus largely on trials and farmer testing of new technology. In another setting, a greater emphasis may be placed on surveys, sociological or information related to the financial picture of the whole farm, and so forth, In the larger picture, none of this matters too much. But it should be remembered that FSD does have certain fundamental characteristics. These include:

- Iteration, such that no single proposal should be considered definitive.
- Integration, usually involving multiple disciplines and explicit building on managerial perspectives and managerial values.
- A consistent orientation towards problem solving and an attitude that no perspective on this problem should be discounted.

FSD also must focus on appropriate linkages with other development support programmes and institutions. In this regard, FSD emphasizes a strongly integrative and integrating philosophy, not just for disciplinary and farming perspectives within FSD itself; hut integrating strengths of personnel, resources, information, and location of the FSD programme with other research and development programmes working in the same area or domain. A critical element of this inter-institutional integration is a logical division of labour for the problem-solving task, This invariably will encourage collaboration in the development of work plans both for short-term as well as medium-and long-term objectives.

FSD can help Identify Appropriate Development Strategies

On-farm systems studies were carried out in village areas south of Bangalore, India to determine the appropriateness of the current livestock policy. Farmers without irrigated crop land are the poorest in this area and at greatest risk from the effects of climatic and biological variations. These farmers depend on small livestock herds for cash income and on subsistence cropping for home consumption. When their small-scale livestock production fails, farmers are forced to seek outside employment for cash, and thus cropping activities suffer.

This farming system study identified two improvements that would contribute to better livestock productivity: control of epizootics and a 'preventative veterinary medicine service' focused on improved fertility management among small herd-owners. Previously, veterinary service policy had been to target larger herds where the immediate payoff had appeared to be much greater. Such a livestock policy change should lead to greater stability in the production of animal products for local markets; less reliance on seasonal employment and, therefore, greater stability of the household labour force; and more stable subsistence cropping activities.

Inter-household Relations and Trade-offs

In Botswana, farmers have a tradition of sharing draught power, In this semi-arid environment, field operations such as ploughing and row planting depend on access to draught in a timely fashion (*i.e.*, days on which there is adequate soil moisture for ploughing/planting). If a technology system is introduced that requires more draught, rather than less, it may mean that the household controlling the draught may not be able to routinely allow other households to also use their animals under a sharing arrangement. Thus, helping one household may harm other households in the community.

Another Intra-generational Issue

An FSD programme in Amphoe Pharao, Thailand discovered a positive and unexpected benefit from interaction at the intra-household level. The introduction of improved and more commercially oriented poultry operations was targeted towards women members of households, because they traditionally manage and derive income from poultry. However, improved technological interventions not only required new inputs by men but were discovered to provide profitable use of under-employed male labour in many households.

This example demonstrates decision making with interactions between members of the household that are desirable for all: a win-win situation. Often, however, decisions for change favour one segment of the household, while adversely impacting on another.

2

Farm Forestry

Farm forestry is the deliberate attempt of farmers towards planting or keeping trees in their farmland in order to supplement daily farm/household needs, meet the contingencies and take market opportunities created by the dwindling supply of tree/forest products from the common lands.

There are reports on trees being increasingly used as contingencies and also contributing to the household food security. Besides, the above points, the tress also contributes to the sustainable land use.

FARM FORESTRY: RESPONSE TOWARDS THE SCARCITY

The real crisis faced by the Third World Countries where firewood, crop residues, kerosene being the popular form of energy for cooking, Eckholm wrote "the people are scrambling to cook dinner under deforestation". However, in the wake of deforestation, villagers coped with increased scarcity in various ways. Introduction of tress in the private land is widely pursued by the farmers elsewhere. A study, using aerial photographs between 1964 and 1988, observed a tremendous increase of private trees in the hills of Nepal. It is postulated that this increase is mainly due to the decreasing access and unavailability of forest resources. This shift of trees has occurred elsewhere. Pakistan gives example of Hazara forests managed by villagers in frontier tribal areas under weak state control. An African study shows " rural households are planting trees in the wake of labour shortages due to off-farm income opportunities.

Farmers favour multipurpose trees over trees that supply only firewood, as cheap substitutes for firewood are available. Similarly planting trees only for fodder alone is not common, but existing fodder trees are retained. Thus introduction of trees in private land is a strategy adopted by villagers to cope with emerging biomass scarcity. Timber trees are stores of values for household savings that do not put demand on household's scarce cash resources. However, tree growing is determined by the farmer's livelihood strategies and resource base. Farmer's decision regarding tree growing depends on:

(1) declining access to the wood land,
(2) loss of communal lands or restrictions on forest access
(3) increasing demand for forest products
(4) increasing demand from the market.

A survey from western Kenya shows that the "poor" prefer firewood. The "average" had a higher portion of fruit and timber while the wealthy invests heavily in fencing. A study on farmer's willingness to grow trees in Gunung Kidul opts for a different reason. As the farmer's willingness to grow trees depends on many factors. An increase in the productivity of staple crop is an important factor permitting farmers to plant trees. Government policy can also create a favourable market trend for growing trees

CONSTRAINS OF TREE GROWING

The following are the main constraints for tree growing by the rural poor:

1. Economic: trees can't compete with other uses for land, labour and capital.
2. Social/cultural: shelter to evil spirits, harbours insects etc.
3. Site: condition not meeting requirement for trees
4. Land use rights tenure security
5. Red tape(ism): Permits for utilisation

The above constraints affect the farmer's willingness to grow trees while Gregersen *et al.*, 1989 gives some reasons with regard to flexibility that can be (dis)incentive for farmers to grow trees:

- Multiple tree outputs (adjust as the market change)
- Flexibility in harvesting time
- Low liquidity in early years reduces flexibility.

Besides the above, the higher rate of discount in case of developing countries discouraged farmers from introducing trees in their farmlands.

Social forestry interventions in south Asia intended to involve local communities, whether it is agro forestry, farm forestry or community forestry needs well planned transformation of a dynamic inter-relationship among community, natural environment and the state. In certain situations, the forest-community dependencies are deep rooted and need careful consideration while designing such development interventions. The theoretical basis for planning such transformations is more complex than project planners and bureaucrats often consider. Most of these forestry interventions went through a number of paradigm shifts, beginning with a top down conventional model approach, to the most recent approach wherein emphasis is laid on local participation. There is a lack of proper understanding by policy makers and foresters, and inadequate theoretical basis in these approaches. Analysis of the approaches may provide explanation for the poor performance of most social forestry projects in south Asia. Different views exist about the poor outcomes and the basic reasons for such outcomes.

Earlier studies that addressed issues of social forestry in south Asia tended to neglect local conceptualisations of natural resources, because their methodologies excluded local histories, local socio-political process and issues of institutional change. Studies in the recent years have started paying attention towards the analysis of local livelihoods, the local complexity of natural resource management and the long traditions of local resource management and conservation (Blaikie and Brookfield 1999; IIED 1998; IDS 1999).

This study will attempt to analyse some of the theoretical contents of the social forestry projects in south Asia and related policies in order to make them explicit and the factors responsible for the poor performance of the social forestry projects with regard to equity, environmental rehabilitation and sustainability.

The present study draws on the authors' research and field experience in social forestry in South Asia (India, Nepal and Sri Lanka) since the mid 90s. Some data was also gathered during specific reviews of three social forestry projects in India, Nepal and Sri Lanka funded by different international donor agencies. The reviews consisted of discussions with local key informants and government officials, several field visits and secondary data collection.

FARM FORESTRY: LAND AVAILABILITY, TAKE-UP RATES AND ECONOMICS

The target set by the Irish Government is to increase the forest cover to 17% of the land area of the country by the year 2030. This involved an annual afforestation level of 25,000 ha to the year 2000 and 20,000 ha thereafter to the year 2030. The target was almost achieved in 1995 but planting levels have fallen far short since, even though substantial financial incentives were put in place.

IMPLICATIONS

Failure to reach the overall target by 2030 will result in failure to reach the critical mass of forest output required to support a viable timber industry. Since the employment creation potential of large timber mills is not very high, it is essential that significant value added accompanies whatever timber output is achieved in order to boost employment. In the absence of achieving the afforestation targets listed in the Strategic Plan for Forestry, the contribution of the forestry sector to the growth targets of the National Development Plan will be less than expected and the objectives set out regarding forestry in the CAP Rural Development Plan will not be achieved.

Background

The study objectives were to examine the factors:

- Influencing land availability for forestry,

- Slowing down or accelerating the rate of afforestation and
- Affecting the economic return from farm forestry.

The amount of land becoming available for afforestation via the market has declined substantially in recent years. Land transfers are via gift or inheritance in the vast majority of cases. Land becoming available (non market) for forestry at time of change of ownership was not examined but a survey of landowners opinions was used to study the factors influencing decisions to plant. In this study the economic returns from forestry were compared with the returns from existing farm enterprises using linear programming techniques. The technique as applied in this study is aimed at maximising total farm gross margin from all available resources thereby arriving at the optimum mix of enterprises (including forestry). The main assumptions of the economic study were as follows:

- Returns to forestry were calculated using a 10% discount rate.
- The gross margin system is used to compare returns from forestry with agricultural enterprises.
- No economic value was given to the amount of carbon dioxide fixed by growing trees, and
- Farm labour was valued both at minimum wage and at average industrial wage rates.

The discount rate normally used by those in the forestry sector is 5%. Fixed costs are proportionally much higher in agriculture than in forestry. Forestry is long-term and all costs are variable in the long term so perhaps a net figure might be more suitable for comparison purposes.

Findings

Land Availability and Take-up Rate

Landholders attitude to forestry are becoming more positive. The unfavourable responses to forestry were based mainly on pragmatic rather than cultural issues, *e.g.* small farm size, land quality, competing EU payments. The main reason for planting was the very favourable income from forestry premium payments on land that had limited other uses. Forestry on good farmland was not favoured but was a definite competitor to agricultural enterprises on land marginal for farming and / or situated some distance away from the main farmyard. Afforestation was mostly associated with large farm size. The study did not examine the question of cutting back in farm enterprises (livestock numbers) which is often associated with old age and / or ill health. The demographic profile of those in the sample with farm forests was the same as in the National Farm situation.

Economic Returns

Opportunities for off-farm employment and levels of remuneration can determine the choice of farm enterprise and the rate of afforestation. Off-farm earnings at or near industrial wage rate levels may result in increases in the area planted often to the exclusion of cattle and sheep enterprises at individual farm level. Since the decision to plant is for all time, economic factors alone are often not the only criteria for decision-making. Where off-farm jobs are not available, forestry will compete with farm enterprises only on farms where there is land surplus to that required to maximise REPS, compensatory allowance, and extensification payments. However, even where no off-farm jobs are available, forestry will always compete with drystock enterprises operated at moderate and low levels of efficiency. The study gives a clear message that a switch to forestry can improve overall farm and household income for many farmers depending on their circumstances. The findings of the study are important pointers for Teagasc staff involved in the Opportunities for Farm Families Programme.

NUTRIENT DYNAMICS IN HIMALAYA

The structure and function of Shisham (Dalbergia sissoo Roxb.) forests were investigated in relation to nutrient dynamics in 5-to 15-year-old stands growing in central Himalaya. Nutrient concentrations and storage in different layers of vegetation were in the order: tree > shrub > herb. Forest soil, litter and vegetation accounted for 80·1-91·9, 1·0-1·5 and 7·0-18·4 %, respectively, of the total nutrients in the system. There were considerable reductions (trees 32·8-43·1; shrubs 26·2-32·4; and herbs 18·8-22·2 %) in nutrient concentrations of leaves during senescence. Nutrient uptake by the vegetation as a whole and also by the different components, with and without adjustment for internal recycling, was investigated. Annual transfer of litter nutrients to the soil from vegetation was 74·8-108·4 kg ha-1 year-1 N, 5·6-8·4 kg ha-1 year-1 P and 38·7-46·9 kg ha-1 year-1 K. Turnover rate and time for different nutrients ranged between 56 and 66 % year-1 and 1·5 and 1·8 years, respectively.

The turnover rate of litter indicates that over 50 % of nutrients in litter on the forest floor are released, which ultimately enhances the productivity of the forest stand. The nutrient use efficiency in Shisham forests ranged from 136 to 143 kg ha-1 year-1 for N, 1441 to 1570 kg ha-1 year-1 for P and 305 to 311 kg ha-1 year-1 for K. Compared with natural oak forest (265 kg ha-1 year-1) and an exotic eucalypt plantation (18 kg ha-1 year-1), a higher proportion of nutrients was retranslocated in Shisham forests, largely because of higher leaf tissue nutrient concentrations. This indicates a lower nutrient use efficiency of Shisham compared with eucalypt and oak. Compartment models for nutrient dynamics have been developed to represent the distribution of nutrients pools and net annual fluxes within the system.

Dry matter accumulation and productivity of forests are influenced by the presence of nutrients in soils and their recycling. Nutrient elements therefore limit forest productivity. Recycling of nutrients is one of the principal processes that supports organic matter production of forests. Thus, the functioning of a forest ecosystem in relation to dry matter production depends not only on the availability of nutrients but also on the pattern and rate of nutrient uptake by species occurring in the forest. According to Lodhiyal and Lodhiyal (1997), survival of tree species and the rate of nutrient uptake depend upon the availability of water.

They pointed out that trees take up large quantities of nutrients from the soil system and, although much of the nutrient uptake is returned to the soil through litter fall, large amount of nutrients are also removed when trees are harvested.

Thus, from the point of view of sustainability, studies on nutrient dynamics in Shisham forests which are harvested following short rotations are important for management of plantation forestry.

Nutrient accumulation and the pattern of distribution in different plants are affected by climate, and by the type and age of the species. Lodhiyal *et al.* (1994) pointed out that the major macronutrients limiting the production of a forest crop are N, P and K.

According to Chapin and Kedrowski (1983), nutrient availability is a major factor influencing the distribution of plant species. Most plants from low nutrient sites have a low nutrient requirement. Stachurski and Zimka (1975) and Tilton (1977) suggested that plants growing on nutritionally deficient sites minimize nutrient loss by retranslocating a greater fraction of N and P from senescing leaves. Uptake of nutrients was found to be much greater in deciduous plantations than in evergreen plantations.

On the basis of above-ground biomass production relative to N uptake, deciduous forests have a lower nutrient use efficiency (NUE) than evergreen forests (Waring and Schlesinger, 1985).

However, NUE did not differ with forest type across local N mineralization gradients in mixed-species sequences, and was poorly correlated with foliar life span across diverse ecosystems.

According to Lodhiyal and Lodhiyal (1997), the high NUE is largely the result of: (a) higher soil nutrient availability, and (b) increased nutrient concentrations and amount of litter nutrient returned through litter fall to the soil. Unscientific thinning and harvesting of trees not only degrades forest soils but also removes nutrients from the ecosystem. Many studies have been carried out into nutrient storage and cycling in forests and plantations. The amounts of nutrients taken up depend on the demands of the plant species and on the availability of nutrients in the soil to supply that demand. Thus, dry matter production is generally based on the fertility of the soil and on extra inputs provided by management.

Increasingly, indigenous leguminous trees are being planted in India because of their high biomass per unit area, their survival on nutrient poor sites and for their nutrient conservation efficiency. The pressures of a growing population and adverse ecological changes in the natural Shisham forests necessitate investigations into their sustainability. Little is known about the nutrient cycling and nutrient use efficiency of Dalbergia sissoo which enable it to occupy Tarai (nutrient-rich sites) belts in central Himalaya. The present investigation aims to provide detailed information on the structure and functioning of Shisham forests growing in moist plains of central Himalayan mountains in India, with regards to nutrient cycling and the pattern of nutrient use.

The major objectives of this study were to describe: (a) the various aspects of nutrient cycling, such as nutrient uptake and nutrient retranslocation in different vegetation layers; (b) nutrient return to soils through litter (*i.e.* by trees, shrubs and herbs); (c) nutrient use efficiency in relation to dry matter production and net nutrient uptake; and (d) to compare estimates of nutrients in Shisham forests with those in exotic plantations and natural forests.

MATERIALS AND METHODS

Nutrient dynamics were investigated in 5-, 10-and 15-year-old Shisham forest stands in Tarai belts of central Himalaya; dry matter dynamics of these forests have been reported previously. Sal forest (mixed broad leaved species) was the original vegetation in the Tarai area. According to Lodhiyal *et al.* (1995a), most of this region was converted into arable land during the 1960s, and since the 1970s has been used for exotic tree plantations.

Details of the study site (*i.e.* topography, geology, soils, meteorological data and vegetation types) and methods of sampling tree, shrub and herb components are described by Lodhiyal *et al.* (2002).

Samples of different tree components, *i.e.* bole wood, bole bark, branch, twig, leaf, reproductive parts, coarse roots (stump root + lateral roots) and fine roots were collected from 12 harvested trees [three trees from each sub-plot, one from each dbh (diameter at breast height) class; for nutrient analysis (nutrient concentration and nutrient content) in each Shisham forest stand.

Tree and shrub components were sampled during September 1996, and herbs (above-and below-ground parts) were collected in each season, *i.e.* summer, winter and the rainy season.

Composite samples (samples of each component taken from the lower, middle and upper plant parts and mixed) for each of the tree, shrub and herb components were taken separately to the laboratory and oven-dried at 60°C to constant weight. Oven-dried samples were mill-ground for nutrient analysis.

Litter samples were collected using litter traps at monthly intervals in all forest stands. The composite litter samples (litter samples collected from each dbh class of tree and mixed to form a composite, separately for each component) for each litter component, *i.e.* leaf, wood, reproductive litter and other litter, were ground separately and analysed for nitrogen, phosphorus and potassium.

Five replicates of dry plant material (0·5 g each) were analysed for total nitrogen after digestion by 10 ml concentrated sulfuric acid using 5 g of a catalyst mixture (potassium sulfate and cupric sulfate in a 9: 1 ratio) in a quick digestion unit. Total nitrogen was determined by the micro-Kjeldahl technique. Phosphorus and potassium were extracted by wet ashing 0·5 g plant material in an acid mixture consisting of 10 ml H_2SO_4 + 3 ml HNO_3 + 1 ml $HClO_4$, following the method of Jackson (1958). Phosphorus was determined by absorption spectrophotometry and potassium by flame photometry.

The standing state of nutrients in trees, shrubs and herbs was computed separately by multiplying the dry weight of each component by the respective nutrient concentration. Nutrient values in trees, shrubs and herbs were summed to obtain the total standing state of nutrients in the vegetation.

The amount of nutrients (N, P and K) in the soil was determined by the micro-Kjeldahl technique for N, absorption spectrophotometry for P and flame photometry for K. The amount of nutrients in the top 30 cm of soil was calculated by summing values for 0-10, 10-20 and 20-30 cm soil depths in each forest stand. Soil volume multiplied by the respective average bulk density gave the weight of the soil, which was in turn multiplied by the corresponding nutrient concentration to obtain the amount of soil nutrients (N, P and K).

The mean nutrient concentration was multiplied by the weight of annual litter fall, *i.e.* leaf, wood, reproductive parts of trees, and leaves of shrubs and herbs to give the amount of nutrients transferred to the forest floor in vegetation in stands of different ages. The turnover rate (K) for each nutrient on the forest floor was estimated following Chaturvedi and Singh (1987b) and Lodhiyal and Lodhiyal (1997) as:

$$K = A/(A + F)$$

where A is the amount of nutrients added to the forest floor by litter fall and F is the nutrient content of the lowest value of standing crop of litter in the annual cycle. Turnover time (t) for nutrients was calculated as for standing litter biomass. Turnover time is the reciprocal of turnover rate (K).

Nutrient uptake by components of vegetation in stands of different age was calculated by multiplying the net primary productivity (NPP) of different components by their nutrient concentration. The values of nutrient uptake by trees, shrubs and herbs were summed to estimate the total annual uptake by the forest vegetation.

Some retranslocation of nutrients (N, P and K) occurred during senescence of foliage. This was assessed following Ralhan and Singh (1987), Rawat and Singh (1988), Lodhiyal *et al.* (1995b) and Lodhiyal and Lodhiyal (1997), *i.e.*

$$R = (X-Y)/X \times 100$$

where R is nutrient retranslocation (%), X is the nutrient mass in mature green leaves, and Y is the nutrient mass in senesced leaves. X and Y were calculated on the basis of nutrient per unit weight of mature green and senesced leaf, respectively, multiplied by total amount of leaf litter fall.

To estimate nutrient retranslocation, 120 mature green and freshly fallen senesced leaves were collected in September (peak month of leaf maturity) and December (leaf senescing period) 1997, respectively. Since rainfall is negligible in the region when leaves senesce (December), leaching is likely to have only a minimal effect on nutrient loss from the leaves. The nutrient use efficiency (E) per unit area of Shisham forests was calculated by:

$$E = P/U$$

where P is NPP and U is the net nutrient uptake in kg ha-1 year-1 by the trees.

RESULTS AND DISCUSSION

Nutrient Concentration

Table below summarizes N, P and K concentrations in the different components of vegetation (*i.e.* trees, shrubs and herbs) of Shisham forests growing in Tarai belts of the central Himalayan mountains in India. The concentration of nutrients in tree components was in the order: reproductive parts > leaf > twig > bole bark > branch > bole wood in above-ground parts, and fine roots > lateral roots > stump root in below-ground parts.

The nutrient concentrations in different components of Shisham trees were higher in 5-year-old stands than in 15-year-old stands. Thus, as trees age the percentage of nutrients (N, P and K) in the different components decreases. Similar trends in different components of fast-growing poplar and eucalypt trees were reported by Lodhiyal *et al.* (1995b), Lodhiyal and Lodhiyal (1997), and Bargali *et al.* (1992) in adjacent areas of central Himalaya.

Table. Concentration of nutrients (% ± s.e.) in different components of trees, shrubs and herbs, and in soils of Tarai Shisham forests in central Himalaya

	Age of Shisham forests (years)			
Components	Nutrient	5	10	15
Tree layer				
Bole wood	N	0·365 ± 0·012	0·340 ± 0·020	0·328 ± 0·014
	P	0·042 ± 0·015	0·039 ± 0·027	0·038 ± 0·018
	K	0·202 ± 0·017	0·198 ± 0·024	0·195 ± 0·010
Bole bark	N	1·410 ± 0·020	1·390 ± 0·025	1·380 ± 0·025
	P	0·089 ± 0·024	0·088 ± 0·020	0·085 ± 0·020
	K	0·588 ± 0·026	0·582 ± 0·020	0·562 ± 0·024
Branch*	N	0·980 ± 0·034	0·962 ± 0·041	0·940 ± 0·052
	P	0·095 ± 0·028	0·088 ± 0·038	0·071 ± 0·012
	K	0·200 ± 0·025	0·198 ± 0·039	0·195 ± 0·014
Twig	N	1·658 ± 0·031	1·625 ± 0·042	1·580 ± 0·025
	P	0·099 ± 0·038	0·096 ± 0·044	0·094 ± 0·020
	K	0·588 ± 0·034	0·582 ± 0·038	0·570 ± 0·025

Leaf	N	2·750 ± 0·055	2·680 ± 0·035	2·640 ± 0·40
	P	0·188 ± 0·059	0·175 ± 0·039	0·170 ± 0·034
	K	0·780 ± 0·062	0·762 ± 0·041	0·740 ± 0·030
Reproductive parts				
	N	3·254 ± 0·048	3·225 ± 0·050	3·118 ± 0·020
	P	0·340 ± 0·043	0·320 ± 0·058	0·312 ± 0·019
	K	0·925 ± 0·040	0·915 ± 0·052	0·908 ± 0·012
Stump root	N	0·570 ± 0·040	0·592 ± 0·030	0·550 ± 0·014
	P	0·042 ± 0·039	0·038 ± 0·024	0·035 ± 0·015
	K	0·378 ± 0·034	0·375 ± 0·025	0·370 ± 0·012
Lateral roots	N	0·610 ± 0·034	0·698 ± 0·034	0·595 ± 0·024
	P	0·068 ± 0·040	0·062 ± 0·032	0·058 ± 0·021
	K	0·550 ± 0·041	0·520 ± 0·028	0·480 ± 0·022
Fine roots	N	1·012 ± 0·055	0·998 ± 0·042	0·995 ± 0·030
	P	0·198 ± 0·059	0·598 ± 0·050	0·188 ± 0·026
	K	0·610 ± 0·059	0·598 ± 0·050	0·589 ± 0·020
Shrub layer				
Lantana camara				
Stem	N	0·705 ± 0·052	0·702 ± 0·184	0·701 ± 0·095
	P	0·079 ± 0·034	0·078 ± 0·098	0·076 ± 0·047
	K	0·510 ± 0·025	0·508 ± 0·075	0·504 ± 0·059
Foliage	N	2·348 ± 0·040	2·345 ± 0·094	2·340 ± 0·052
	P	0·155 ± 0·015	0·152 ± 0·084	0·150 ± 0·038
	K	0·738 ± 0·021	0·735 ± 0·070	0·730 ± 0·044
Roots	N	0·780 ± 0·058	0·778 ± 0·062	0·770 ± 0·074
	P	0·077 ± 0·028	0·075 ± 0·048	0·072 ± 0·046
	K	0·546 ± 0·036	0·540 ± 0·042	0·540 ± 0·070
Murraya koenighii				
Stem	N	0·675 ± 0·069	0·674 ± 0·084	0·670 ± 0·053
	P	0·047 ± 0·022	0·045 ± 0·046	0·044 ± 0·058
	K	0·350 ± 0·062	0·348 ± 0·078	0·340 ± 0·063
Foliage	N	1·252 ± 0·042	1·250 ± 0·043	1·246 ± 0·047
	P	0·118 ± 0·048	0·118 ± 0·034	0·118 ± 0·064
	K	0·750 ± 0·018	0·750 ± 0·048	0·746 ± 0·072
Roots	N	0·758 ± 0·018	0·755 ± 0·024	0·752 ± 0·034
	P	0·049 ± 0·027	0·046 ± 0·057	0·045 ± 0·094
	K	0·412 ± 0·044	0·410 ± 0·054	0·408 ± 0·050
Clerodendron viscosum				
Stem	N	0·642 ± 0·075	0·640 ± 0·046	0·638 ± 0·029
	P	0.051 ± 0·081	0·050 ± 0·054	0·050 ± 0·039
	K	0·335 ± 0·048	0·332 ± 0·070	0·350 ± 0·066
Foliage	N	1·158 ± 0·054	1·150 ± 0·028	1·152 ± 0·045
	P	0·130 ± 0·029	0·128 ± 0·038	0·127 ± 0·050
	K	0·678 ± 0·052	0·675 ± 0·056	0·673 ± 0·043

Roots	N	0·670 ± 0·035	0·667 ± 0·037	0·664 ± 0·032
	P	0·055 ± 0·024	0·052 ± 0·042	0·051 ± 0·047
	K	0·358 ± 0·054	0·352 ± 0·042	0·350 ± 0·052
Pogostemone benghalense				
Stem	N	0·681 ± 0·082	0·680 ± 0·094	0.675 ± 0.074
	P	0·054 ± 0·050	0·050 ± 0·045	0.046 ± 0.058
	K	0·342 ± 0·042	0·340 ± 0·072	0.334 ± 0.073
Foliage	N	1·180 ± 0·053	1·178 ± 0·057	1·175 ± 0·046
	P	0·139 ± 0·042	0·136 ± 0·028	0·135 ± 0·034
	K	0·684 ± 0·038	0·682 ± 0·065	0·677 ± 0·054
Roots	N	0·750 ± 0·042	0·740 ± 0·069	0·730 ± 0·063
	P	0·055 ± 0·034	0·054 ± 0·047	0·052 ± 0·050
	K	0·480 ± 0·026	0·479 ± 0·046	0·475 ± 0·042
Herb layer				
Above-ground				
	N	1·572 ± 0·068	1·540 ± 0·076	1·475 ± 0·082
	P	0·084 ± 0·049	0·082 ± 0·056	0·078 ± 0·044
	K	1·342 ± 0·063	1·314 ± 0·074	1·304 ± 0·064
Below-ground				
	N	1·221 ± 0·063	1·202 ± 0·058	1·165 ± 0·078
	P	0·065 ± 0·025	0·062 ± 0·042	0·060 ± 0·036
	K	1·142 ± 0·052	1·135 ± 0·063	1·124 ± 0·050
Soil (0-30 cm depth)				
	N	0·159 ± 0·30	0·156 ± 0·32	0·151 ± 0·39
	P	0·013 ± 0·26	0·013 ± 0·26	0·012 ± 0·39
	K	0·065 ± 0·38	0·061 ± 0·30	0·056 ± 0·40

* Large woody shoots with leaves.

Current shoots bearing leaves.

Main root with tap roots and parts of stem (stump bearing roots).

Lateral branches of main roots with diameter > 5 mm.

Roots of diameter < 5 mm and associated mycorrhizae.

Leaf to wood concentration ratios for N (7·5-8·0), P (4·5) and K (3·8-3·9) were similar to those found in natural forests of central Himalaya and fall within the range for N (11-12) and P (6-7) reported for poplar trees but are much lower than values reported for temperate forests.

Leaf nutrient concentrations are higher than in Populus deltoides plantations and Eucalyptus hybrid plantation.

Soil nutrient concentration decreased with increasing forest age, possibly as a consequence of uptake and leaching of nutrients by the Shisham forests. Similar findings were also reported for Tarai poplar plantations in nearby areas.

However, loss of nutrients can be considerable during initial site preparation, which involves burning ground vegetation and litter and, as a result of soil disturbance which accompanies planting, especially when vegetation cover is sparse.

STANDING STATE OF NUTRIENTS

The standing state of nutrients increased with forest age due to increased dry matter storage. Bole (wood + bark) had the greatest concentration of nutrients in the trees. The relative contribution to the standing state of nutrients in different above-ground parts was generally in the order: bole > leaf > branch > reproductive parts, and that in below-ground parts was: stump root > lateral roots > fine roots. Similar trends have been reported for eucalypt and poplar plantations. The different nutrients in the total vegetation (trees + shrub + herbs) was in the order: N (488-1002 kg ha-1 year-1) > K (235-471 kg ha-1 year-1) > P (50-82 kg ha-1 year-1).

Table. Nutrient content (kg ha-1) in different vegetation layers of Shisham forests

	Age of Shisham forests (years)		
Components	5	10	15
Nitrogen Tree layer	402·1 (82·3)	692·2 (85·5)	877·3 (87·5)
% Allocation in bole*	43·4	45·3	45·0
branch	17·0	18·7	20·3
leaf	22·1	18·1	15·5
reproductive parts	4·3	4·4	5·5
coarse roots	11·3	11·3	11·2
fine roots	1·9	2·2	2·5
Shrub layer	44·4(9·1)	79·8 (9·8)	91·3 (9·1)
Herb layer	41·9 (8·6)	38·1 (4·7)	33·6 (3·4)
Total vegetation	488·4 (100)	810·1(100)	1002·2 (100)
Soil (0-30 cm depth)	4834·6	4945·4	5026·0
Phosphorus Tree layer	43·7 (87·8)	64·5 (88·3)	72·5 (88·7)
% Allocation in bole*	37·1	44·8	48·6
branch	12·6	15·5	16·8
leaf	13·9	12·7	12·0
reproductive parts	24·3	4·7	6·6
coarse roots	8·7	9·6	10·2
fine roots	3·4	12·7	5·8
Shrub layer	3·9 (7·8)	6·6(9·0)	7·4 (9·1)
Herb layer	2·2 (4·4)	2·0 (2·7)	1·8 (2·2)
Total vegetation	49·8 (100)	73·1 (100)	81·7 (100)
Soil (0-30 cm depth)	405·7	412·6	388·9
Potassium Tree layer	172·6 (73·6)	289·6 (79·0)	390·9 (83·1)
% Allocation in bole*	50·0	55·4	51·9
branch	10·7	6·4	12·6

leaf	14·6	12·3	9·7
reproductive parts	2·8	3·0	3·6
coarse roots	19·2	19·8	18·8
fine roots	2·7	3·1	3·4
Shrub layer	25·6(10·9)	43·7 (11·9)	49·5 (10·5)
Herb layer	36·4 (15·5)	33·2 (9·1)	30·3 (6·4)
Total vegetation	234·6 (100)	366·5 (100)	470·7 (100)
Soil (0-30 cm depth)	1968·4	1934·6	1872·1

Values in parentheses represent percentage contribution.
* Bole wood + bole bark, which accounted for 11·0-19·8 % of the values (N, P and K). Branch + twig (new shoots bearing leaves), which accounted for 1·7-9·3 % of the values (N, P and K). Stump root (main root) + lateral roots (lateral branches of main roots, which accounted for 3·2-7·9 % of the values (N, P and K).}

Nutrient storage by shrubs accounted for 8-12 % of storage by the vegetation as a whole and increased with forest age, while nutrient storage by herbs (2-15 %) decreased with stand age.

Similar trends were also reported for fast-growing poplar plantations. This is because the tree canopy affects the growth of herbs and also reduces the amount of light available and thus photosynthetic efficiency.

Nutrient Retranslocation

The retranslocation of nutrients from senesced leaves was in the order: K (40-43 %) > N (33-35 %) > P (28-30 %). These estimates of nutrient retranslocation are markedly higher than those reported for fast-growing exotic eucalypt plantations.

However, the present estimates are lower than values reported for Tarai poplar plantations (42-64 %; Lodhiyal *et al.*, 1995b), for Larix laricina, Alnus crispa and Betula papyrifera (44-81 %; Chapin and Kedrowski, 1983), for Populus tremuloides (42-65 %; Verry and Timmons, 1976) and Populus deltoides (74-89 %; Baker and Blackman, 1977).

However, our estimates are close to the 23-39 % reported for Quercus rubra (Grizzard *et al.*, 1976) and 38-49 % for Betula alleghaniensis. The estimate of N retranslocation was 33 % in mixed northern hardwood species, which is similar to the present N estimate.

Thus, on the basis of correlation between proportional nutrient retranslocation from senesced leaves and tree leaf nutrient concentrations, our findings are consistent with those of Chapin and Kedrowski (1983).

According to Lodhiyal and Lodhiyal (1997), the nutrient requirement of vegetation not only depends on stand age and soil nutrient level, but also on the nutrient concentration of foliage.

They noted that the greater the leaf tissue nutrient level, the higher the nutrient retranslocation capacity of forest trees would be.

Table. Magnitude of retranslocation of nutrient (%) in leaves of different vegetations layers of Shisham forests

	Age of Shisham forests (years)			
Vegetation	Nutrient	5	10	15
Tree layer	N	35·2	33·6	32·8
	P	30·3	28·8	28·1
	K	43·1	41·5	40·3
Shrub layer	N	32·4	30·3	28·8
	P	30·7	25·0	25·0
	K	30·2	28·7	26·2
Herb layer	N	22·2	21·6	20·8
	P	17·2	16·8	16·9
	K	19·8	19·4	18·8

Nutrient Uptake

The present estimates of gross nutrient uptake (kg ha-1 year-1) by vegetation ranged from 114 (5-year-old stand) to 167 (15-year-old stand) for N, 10 (5-year-old stand) to 14 (15-year-old stand) for P, and 70 (5-year-old stand) to 92 (15-year-old stand) for K. Of this, the tree layer accounted for 59-74 % N, 64-81 % P and 44-61 % K; the shrub layer for 5-6 % N, 6-14 % P and 4-6 % K; and the herb layer for 20-37 % N, 13-22 % P and 33-52 % K.

Table. Nutrient uptake (kg ha-1 year-1) by different vegetation layers of Shisham forests

	Age of Shisham forests (years)		
Components	5	10	15
Nitrogen			
Tree layer	67·1(63·7)	96·9 (91·5)	122·9(117·1)
% Allocation in			
bole*	45·1	41·3	40·4
branch	17·7	17·1	18·3
leaf	14·4	16·5	14·4
reproductive parts	3·0	7·7	6·6
coarse roots	11·7	9·2	12·2
fine roots	8·1	8·2	8·1
Shrub layer	5·4 (4·9)	9·3 (8·6)	10·2 (9·4)
Herb layer	41·9 (34·3)	38·1 (31·4)	33·6 (28·0)
Total vegetation	114·4 (102·9)	144·3 (131·5)	166·7 154·5)
Phosphorus			
Tree layer	6·4 (6·2)	8·9 (8·7)	10·9 (10·6)
% Allocation in			
bole*	44·1	41·2	40·3

branch	14·9	14·2	14·0
leaf	10·4	11·6	10·4
reproductive parts	3·3	8·2	7·4
coarse roots	10·5	7·4	10·9
fine roots	16·8	17·4	17·0
Shrub layer	1·4 (1·3)	0·75 (0·7)	0·8 (0·7)
Herb layer	2·2 (1·9)	2·0 (1·7)	1·8 (1·5)
Total vegetation	10·0 (9·4)	11·7 (11·1)	13·5 (12·8)
Potassium			
Tree layer	30·5 (29·3)	42·8 (40·9)	56.1 (54·0)
% Allocation in			
bole*	49·0	47·9	45·6
branch	10·5	10·6	11·1
leaf	9·0	10·6	8·8
reproductive parts	2·0	4·9	4·2
coarse roots	18·7	14·8	19·8
fine roots	10·8	11·2	10·5
Shrub layer	3·1 (2·9)	5·1(4·7)	5·5 (5·2)
Herb layer	36·4 (30·5)	33·1 (28·0)	30·3 (25·8)
Total vegetation	70·0 (62·7)	81·0 (73·6)	91·9 (85·1)

Values in parentheses represent nutrient uptake (kg ha-1 year-1) after adjustment for retranslocation.

* Bole wood + bole bark, which accounted for 11·9-19·6 % of the nutrient values (N, P and K).

Branch + twig (new shoots bearing leaves, which accounted for 4·9-8·3 % of the nutrient values (N, P and K).

Stump root (main root) + lateral roots (lateral branches of main root, which accounted for 2·2-10·3 % of the nutrient values (N, P and K).

Nutrient Return through Litter Fall

Of the net annual uptake of nutrients by trees, 56-69 % N, 63-72 % P and 31-46 % K is returned to the ground through litter fall. The total inputs of nutrients to the forest floor via tree litter fall and litter fall from the total vegetation were similar to those of a eucalypt plantation (56 % N, 51 % P and 47 % K; Bargali and Singh, 1991) and lower than those reported for Populus deltoides plantations in an adjacent area in the central Himalayan mountains.

Table. Mean nutrient concentration (% ± s.e.) in different litter components of vegetation of Shisham forests in Tarai of central Himalaya

	Age of Shisham forests (years)		
Components	5	10	15
Nitrogen			
Trees			
Leaf	1·783 ± 0·088	1·779 ± 0·084	1·774 ± 0·085

Wood	1·011 ± 0·050	0·982 ± 0·052	0·954 ± 0·051
Reproductive parts			
	1·952 ± 0·068	1·935 ± 0·078	1·870 ± 0·080
Shrubs*	1·003 ± 0·092	1·032 ± 0·090	1·055 ± 0·088
Herbs	1·572 ± 0·068	1·540 ± 0·076	1·475 ± 0·082
Phosphorus			
Tree			
Leaf	0·131 ± 0·042	0·124 ± 0·040	0·122 ± 0·045
Wood	0·070 ± 0·048	0·067 ± 0·046	0·066 ± 0·044
Reproductive parts			
	0·221 ± 0·068	0·208 ± 0·062	0·203 ± 0·064
Shrubs*	0·093 ± 0·068	0·099 ± 0·066	0·099 ± 0·067
Herbs	0·084 ± 0·049	0·082 ± 0·060	0·078 ± 0·044
Potassium			
Trees			
Leaf	0·443 ± 0·062	0·445 ± 0·056	0·440 ± 0·054
Wood	0·395 ± 0·070	0·390 ± 0·062	0·382 ± 0·066
Reproductive parts			
	0·509 ± 0·070	0·503 ± 0·068	0·499 ± 0·066
Shrubs*	0·497 ± 0·058	0·506 ± 0·054	0·521 ± 0·056
Herbs	1·342 ± 0·063	1·314 ± 0·074	1·304 ± 0·064

* This includes the composite mean nutrient concentrations (%) of leaves, twigs and reproductive parts of shrubs.

This includes the (%) nutrient concentration of leaves of herbs.

Table. Amount of nutrients (kg ha-1 year-1) returned through different litter components of vegetation

	Age of Shisham forests (years)		
Components	5	10	15
Nitrogen			
Trees	41·9 (56·0)	64·1 (65·6)	74·9 (69·0)
% Allocation in tree component			
Leaf	48·6	55·4	56·2
Wood	1·2	2·1	3·2
Reproductive parts	0·7	1·9	3·2
Roots	5·6	6·2	6·3
Shrubs	5·8 (7·8)	9·1 (9·3)	12·3 (11·3)
Herbs	27·0 (36·2)	24·5 (25·1)	21·3 (19·6)
Total vegetation	74·8 (100)	97·6 (100)	108·4 (100)
Phosphorus			
Trees	3·6 (63·3)	5·3 (70·1)	6·1 (72·4)

% Allocation in tree component			
Leaf	46·9	50·1	49·8
Wood	1·1	1·8	2·9
Reproductive parts	1·1	2·5	4·4
Roots	14·2	15·7	15·3
Shrubs	0·5 (9·5)	0·9 (11·6)	1·1 (13·6)
Herbs	1·5 (27·2)	1·4 (18·3)	1·2 (14·0)
Total vegetation	5·6 (100)	7·5 (100)	8·4 (100)
Potassium			
Trees	12·0 (31·0)	18·5 (41·5)	21·5 (45·9)
% Allocation in tree component			
Leaf	23·3	30·4	32·4
Wood	0·9	1·9	3·0
Reproductive parts	0·4	1·0	1·9
Roots	6·4	8·2	8·6
Shrubs	2·9 (7·5)	4·4 (10·0)	6·0 (13·0)
Herbs	23·8 (61·5)	21·5 (48·5)	19·3 (41·1)
Total vegetation	8·7 (100)	44·5 (100)	46·9 (100)

Values in parentheses are the percent contribution of the total vegetation.

Turnover of Nutrients on the Forest Floor

The turnover time of nutrients on the forest floor was longer than that of tropical dry deciduous forests (0·3-0·5 year; Pandey, 1980), eucalypt plantations and Populus deltoides plant ations.

This indicates that a small part of the forest floor litter biomass remained in steady state but this amount was less than in other deciduous forests.

According to Lodhiyal and Lodhiyal (1997), a long turnover time shows that the forest floor litter biomass is in steady state because the annual decomposition rate of litter is low compared with the rate of litter input.

Table. Turnover rate (K, year-1) and turnover time (t, years) of nutrients on the forests floor

	Age of Shisham forests (years)			
Nutrient		5	10	15
N	K	0·65	0·66	0·63
	t	1·54	1·51	1·59
P	K	0·62	0·65	0·56
	t	1·61	1·54	1·79
K	K	0·64	0·66	0·61
	t	1·56	1·52	1·64

NUTRIENT USE EFFICIENCY

The nutrient use efficiency (NUE = NPP /net nutrient uptake) of Shisham trees ranged from 137-143 kg ha-1 year-1 for N, 1441-1570 kg ha-1 year-1 for P and 305-311 kg ha-1 year-1 for K. According to Lodhiyal *et al.* (1995b), as stems and branches become larger the nutrient concentration of woody tissues declines, resulting in higher nutrient use efficiency. The availability of soil nutrients in the Shisham forests studied was found to be similar to that in nearby high and low density Populus deltoides plantations. This may be due to higher nutrient concentrations in tissues and increased amounts of nutrients returned through litter in Shisham forests.

Table : Nutrient use efficiency (kg ha-1 year-1) of Tarai Shisham forests in central Himalaya

	Age of Shisham forests (years)		
Nutrient	5	10	15
N	143 (113)	137 (111)	143 (112)
P	1484 (1177)	1441 (1170)	1570 (1228)
K	311 (247)	305 (240)	309 (242)

NUE = NPP/net nutrient uptake.
Values in parentheses are above-ground NUE.

Nitrogen NUE for N was lower in Shisham forests than in a nearby poplar plantation while NUE for P and K was similar in both sites.

However, the NUE for N was higher than the value reported for an 8-year-old eucalypt stand (215 kg ha-1 year-1; Bargali and Singh, 1991) and 189-209 kg ha-1 year-1 for central Himalayan natural Sal (Shorea robusta), oak (Quercus leucotrichophora) and chir (Pinus roxburghii) forests.

This indicates that Shisham trees are preferable from the standpoint of NUE and their fast growth rate. Despite a proportional increase in nutrient retranslocation, Shisham forests have a similar NUE to eucalypt stands, largely because of higher tissue nutrient concentrations.

The soil nutrient extraction efficiency (nutrient uptake per unit nutrient present in soil) of the Shisham forests studied was 1·3-2·2 for N, 1·5-2·7 for P and 1·5-2·9 for K, and falls within the range reported for Populus deltoides plantations.

On the basis of the nutrient parameters above, Dalbergia sisoo is the most important species because it not only produces a higher amount of dry matter but also because it requires only small amounts of nutrients compared with exotic tree plantations and indigenous forest species in adjacent areas.

Thus, it is more suitable than other species not only for Tarai regions but also for plantations in areas of low nutrient status.

Thus, it is concluded that growing Shisham forests in the Tarai belt is advisable because of their relatively higher productivity, even in poor sites, than any other natural forest growing in adjacent areas. In addition, the species removes only small amounts of nutrients compared with the exotic fast-growing trees (poplar and eucalypt). Therefore, trees can be harvested in a shorter rotation period with higher wood biomass in this moist plain of central Himalaya.

NUTRIENT DYNAMICS

Compartment models of nutrient dynamics (pools and fluxes) for 5-and 15-year-old Shisham forests. Nutrients present in the soil to a depth of 30 cm are considered a source, while those associated with decomposition are released into the soil for re-use.

The direction of nutrient flux from soil to foliage indicates a one-way movement, although it is realised that the nutrients are utilized by the leaf in organic matter synthesis and that they are redistributed among different components at varying rates giving rise to internal recycling.

The amount of nutrients (N, P and K) stored in the vegetation was greater in 15-year-old than in 5-year-old forest stands. Of the total nutrients, the tree layer accounted for 74-88 % in the 5-year-old stand and 83-89 % in the 15-year-old stand. The total amount of P and K in the soil decreased, while the amount of N increased with increasing age of the forest.

The amount of nutrients (N, P and K) taken up by the different vegetational layers as a proportion of the total uptake was as follows: 5-year-old forest, trees (47-65 %) > herbs (20-49 %) > shrubs (5-14 %); 15-year-old forest, trees (64-82 %) > herbs (12-30 %) > shrubs (5-6 %).

The total amount of nutrients retranslocated from above-ground senescing plant parts of the tree layer increased with the age of forest from 3·4 (5-year-old stand) to 5·8 kg ha-1 year-1 (15-year-old stand) for N, 0·2 (5-year-old stand) to 0·3 kg ha-1 year-1 (15-year-old stand) for P, and 1·2 (5-year-old stand) to 2·0 kg ha-1 year-1 (15-year-old stand) for K. In the shrub layer, the amount of N, P and K retranslocated was 0·37-0·84, 0·04-0·06 and 0·13-0·34 kg ha-1 year-1, respectively, and increased with the age of the stand.

However, in the herb layer, the amount of nutrients ranged from 5·6 to 7·7 for N, 0·2 to 0·3 for P and 4·5 to 5·9 kg ha-1 year-1 for K, and decreased with the age of the stand. Thus, a proportion of annual dry matter production of litter supports the whole pattern of nutrient transfer in Shisham forests and consequently reduces the nutrient demand on the soil system.

3

Land Drainage Improvements and Land Policy

LAND DRAINAGE PROJECTS

Land drainage projects fall of the Directive. New land drainage works, including flood defence works and defences against the sea, require planning permission and the provisions described in Parts 1 and 2 of this booklet apply to such works. Land drainage improvement works undertaken by drainage bodies are permitted development under the Town and Country Planning (General Permitted Development) Order 1995 and are therefore exempt from planning permission. As such works might have significant effects on the environment, the principles of EIA need to be applied to them. This is done through the Environmental Impact Assessment (Land Drainage Improvement Works) Regulations 1999, which apply to England and Wales and revoke and replace earlier Regulations which came into force in 1988 and were amended in 1995.

The 1999 Regulations require a drainage body to consider whether proposed improvement works are likely to have significant effects on the environment. Where the drainage body considers that there are unlikely to be such effects, it must publicise its intention to carry out the works. If the drainage body receives representations that there are likely to be significant effects but it still thinks otherwise, it must apply for a determination to the appropriate authority (Secretary of State for the Environment, Food and Rural Affairs or, in Wales, the National Assembly for Wales).

Where the drainage body concludes that the works are likely to have significant environmental effects, it must publicise its intention to prepare an environmental statement and notify specified consultation bodies (English Nature, the Countryside Agency and any other authority or organisation which might have an interest).The environmental effects of the improvement works must be assessed in the light, in particular, of the environmental statement and any representations received. If there are no objections, the drainage body may determine that it will proceed with the works. If there are objections, the proposal must be referred to the appropriate authority for a determination giving or refusing consent to the works. The determination must be publicised.

PORTS AND HARBOURS

The Directive includes trading ports, piers for loading and unloading connected to land and outside ports (excluding ferry piers) which can take vessels of over 1350 tonnes. Includes construction of harbours and port installations, including fishing harbours, which do not fall.

Proposed development at ports and harbours down to the low water mark is subject to Town and Country Planning EIA Regulations. For other works, the requirements of the Directive have been implemented in England, Wales and Scotland by the Harbour Works (Environmental Impact Assessment) Regulations 1999. These Regulations incorporate a revised Schedule 3 to the Harbour Works Act 1964, which gives the procedures for harbour revision orders and harbour empowerment orders, and replace the Harbour Works (Assessment of Environmental Effects) (No 2) Regulations 1989 governing harbour works not needing to be authorised by harbour order.

Whether or not proposed works are subject to a harbour order, the appropriate Minister (generally the Secretary of State for Transport) is responsible for deciding whether EIA is required and, if so, whether the works should proceed in the light of the assessment and the comments of environmental bodies.

In general, works under harbour empowerment orders are likely to fall the Directive and require EIA in every case, whereas works under harbour revision orders are likely to fall and the need for EIA will depend on whether there are likely to be significant environmental effects. Proposed works which are subject to harbour orders are not subject to EIA if the area of works comprised in the project does not exceed 1 hectare, unless any part falls within a `sensitive area' or the Secretary of State decides that EIA is needed in a particular case.

Marine Fish Farming

Intensive fish farming is project. Developments of onshore fish farming facilities may require planning permission and the provisions described in Parts 1 and 2 of this booklet apply to proposals for such facilities.

Off-shore facilities do not require planning permission but require a lease from the Crown Estate Commissioners (or, where appropriate, from the Shetlands Islands Council or Orkney Islands Council). The Environmental Impact Assessment (Fish Farming in Marine Waters) Regulations 1999 (SI No 367), which apply to England, Wales and Scotland, set indicative criteria which, when triggered, require the Commissioners to screen applications to determine whether EIA is needed. They apply where the development is designed to hold a biomass of 100 tonnes or more, or where it will extend to 0.1 hectare or more of the surface area of the marine waters (including any proposed structures or excavations), or where any part of the development is to be carried out in a sensitive area. The Regulations prohibit a lease from being granted before the environmental statement provided by the developer, and comments on it from consultees and the general public, have been considered.

The Crown Estate Commissioners have, in consultation with the Scottish Executive Rural Affairs Department, issued an Environmental Assessment Guidance Manual for Marine Salmon Farmers which provides guidance to developers on how to conduct EIA and prepare an environmental statement. Guidance on the practical implications of the 1999 Regulations has also been issued by the Scottish Executive Rural Affairs Department.

Marine Dredging for Minerals

Extraction of minerals by marine dredging falls of the Directive. The Environmental Impact Assessment and Habitats (Extraction of Minerals by Marine Dredging) Regulations, which are likely to come into force early in 2001, will implement the Directive in waters around the UK (but excluding waters around Scotland). They will also implement the Habitats Directive (92/43/EEC).

Unless the regulator (Secretary of State in respect of English Waters or National Assembly in respect of Welsh waters) has provided, in response to a request, a written determination that proposed dredging is not likely to have significant effects, application for permission must be obtained and the application must be accompanied by an environmental statement. The only exceptions to this are where the Secretary of State has determined that the dredging would form part of a project serving national defence purposes, or where the regulator has determined that it is exempted under Article 2.3 of the Directive.

Transport and Works Act 1992

The Transport and Works Act 1992 (TWA) enables orders to be made authorising the construction or operation of railways, tramways, other guided transport systems and inland waterways; and works interfering with rights of navigation. Orders may also authorise ancillary matters such as compulsory purchase and creating or extinguishing rights over land. Applicants for TWA orders may apply at the same time for a direction that planning permission is deemed to be granted.

The procedures for making applications for orders under Part 1 of the Act are contained in the Transport and Works (Applications and Objections Procedure) (England and Wales) Rules 2000 (SI No 2190), which also implement the Directive. These Rules apply in England and Wales and replace the earlier

1992 Rules (SI No 2902) except for applications made and not determined before the new Rules come into force on 16 October 2000. In England, most applications for orders are determined by the First Secretary of State. Applications for orders relating wholly to Wales fall to be determined by the National Assembly for Wales.

The procedures include provision for opposed orders to be considered by way of a public inquiry, hearing or exchanges of written representations. Schemes that are identified by the Secretary of State or the Assembly as being of national significance are also subject to Parliamentary approval prior to a public inquiry being held.

LAND POLICY IN URBAN PLANNING

Since urban planning in India is largely concerned with development of land, it would be relevant to briefly consider how perceptions about land and real estate property have evolved. The Indian Constitution initially recognised 'to acquire, hold and dispose of property' as a fundamental right. Consequently when land was to be compulsorily acquired 'compensation' at market price was payable. Subsequently the term compensation was replaced by the term 'amount'. This ideology culminated in the enactment of Urban Land (Ceiling and Regulation) Act 1976 that attempted nationalisation of vacant urban land by paying nominal amount. Finally the fundamental right to property was deleted from the Constitution. The first articulation of the Urban Land Policy was proposed by the Urban Land Policy Committee (Ministry of Health) appointed by the Government of India in 1965. The Committee articulated the following Land Policy Objectives

1. To achieve optimum social use of urban land;
2. To make land available in adequate quantity, at right time and for reasonable prices to both public authorities and individuals;
3. To encourage cooperative community effort and bona fide individual builders in the field of land development, housing and construction;
4. To prevent concentration of land ownership in a few private hands and especially to safeguard the interests of the poor and under-privileged sections of the urban society.

Further the Committee observed that to realise the objectives "there is no escape from large scale public acquisition if the question of guiding urban development or the provision of adequate housing and other facilities is to be tackled effectively and large scale advance acquisition of land would really be in the interests of the society as a whole. It is by far the best and perhaps the only way to put an end to speculation in land and to capture subsequent increases in land values. These surpluses, where realised by the public authorities, should benefit the community in more ways than one." Not surprising the role models of Indian Town Planners — Delhi Master Plan,

Chandigarh, Gandhinagar and Navi Mumbai were all based on public ownership of land. Whether public ownership in fact achieved the land policy objectives in such cases may be a matter of debate. But a verdict on Delhi experience was;

1. It has not been possible for DDA to provide land at affordable prices to low income beneficiaries resulting in large scale jhuggi jhopadi colonies.
2. In the absence of price signals land has been sub optimally used, resulting in over provision to powerful groups, and
3. DDA's policy to auction very few plots at a time and treating the maximum price quoted in such biding as the real market price has in fact meant artificially increasing the land price through deliberate scarcity."

However, securing large-scale public ownership of land implied compulsory acquisition of land. There was considerable discontent amongst the original landowners about the manner in which compensation was determined and paid. The Land Acquisition Act 1894 initially provided the date of declaration of intention to acquire the land as the reference date for determining the market value. However no time limit was laid down for actual payment of compensation. 1984 amendments introduced the time limit of three years and also provided for payment of interest from the date of award to actual payment or possession of land and solatium of 30% of market value. However the market value is to be reckoned at current use value at the exclusion of expected rise in value on account of future use. The proposed changes in the LA Act and the R and R Policy attempt to remove many of these lacunae. But planned urban development is not being recognised as a public purpose for which powers of eminent domain could be used and in practical terms the proposed method of deciding compensation and rehabilitation package would make recourse to compulsory acquisition of land expensive for lands that also require substantial investment in trunk infrastructure. This would compel search for new paradigm in respect of urban land. The thinkers in the first world too were enamoured by the socialistic notion of community ownership of land.

However on practical considerations they sought solutions short of nationalisation of urban land. The two extreme proposals had one thing in common. Both considered right to own land and right to develop (or build upon) as separate rights. Henry George argued that a private landowner may have right to own both land and development rights. But he has no rights on the rents accruing to land, as they are results of monopoly and not efforts of the owner. He therefore argued that the state has legitimate right to recover 100% of such rents by way of taxes. He allowed the owner to retain the returns on his investments in improvements. He was also prophetic about the ills of public ownership of land.

In 1879 he stated, "I do not propose to purchase or confiscate private property in land. It is not necessary to confiscate land — only to confiscate rent. Taking rent for public use does not require that the state lease land; that would risk favouritism, collusion, and corruption." On the other hand a committee under the chairmanship of Justice Uthwatt in UK suggested that the betterment occurring on account of development of land should be balanced with the compensation to be paid for acquiring land for public purposes. To enable recovery of betterment the Committee proposed nationalisation of development rights (by paying compensation).

Development required planning permission subject to payment of betterment. Despite three attempts to recover betterment since 1947 it has still not succeeded. A cryptic comment on this reads, "The state expropriates property rights, and then charges those from whom it has taken those rights for granting permission to use them on its terms. A betterment levy is wrong in principle, and like most things that are fundamentally wrong, it will always fail in practice" The Indian urban planning thought under the pre 1991 macroeconomic framework was oblivious of property rights and resultant land and real estate market. Hernando de Sotto argued the importance of clarity of property rights and labelled poorly recorded property rights as the 'dead capital' unable to ensure finance for poor. But his arguments have evoked little debate in India. Lack of conceptual clarity about land and property rights have given rise to many expedient policy initiatives.

Instrument of Nationalising Development Rights

FAR essentially a zoning tool in US cities rationalises the intensity of development that could be permitted considering the existing level of development, accessibility and use. The FAR in US cities varies considerable from less than one to 15. However in most Indian cities such considerations are not used in defining FAR. In many states common FSI values are prescribed across all cities as part of state wide building regulations and do not form part of master plan of individual city.

There has been two-fold argument justifying this position. First, varying FSI within a city would be seen as discriminatory between different land owners and second, varying FSI across cities would lead to demand from political quarters to increase rationally defined FSI to an arbitrary level proposed in another city. The general tendency has been to prescribe low uniform FSI (around one). This has meant scarcity of development rights particularly in cities that experience faster economic and population growth and resultant increase in demand for per capita floor space. Instead of adopting measures to reduce scarcity of development rights by rationalising FSI pattern, retaining existing low FSI regime is being implicitly used as an instrument of nationalising development rights beyond prescribed FSI (without paying compensation).

Armed with such nationalised development rights the state administrators could allot these rights on conditions of payment or fulfilling other obligations like providing free houses to slum dwellers. (Hyderabad Master Plan proposed a base FSI and permissible increase subject to payment. Maharashtra Government increased FSI in Mumbai subject to payment that would be equally shared between Municipal Corporation and the state government. Extra FSI is similarly allowed in Chennai. Similarly extra FSI is allowed for rehabilitating existing occupants of rent controlled buildings and slums, free of cost). In this context the objectives of raising revenue or helping a class (not necessarily poor) will succeed when base FSI is low and scarcity of development rights is created. Creating scarcity is not a healthy way of managing any market.

FINDINGS AND SUGGESTIONS OF URBAN PLANNING

Urban planning is basically concerned with the location, intensity and amount of land development required for various space using functions of city life industry, wholesaling, business, housing, recreation, education, religious and cultural activities of the people. Thus urban planning is itself neutral. But the institutionalisation of planning practice within a complex bureaucracy has contributed to the re-politicisation of urban planning. It has become a mode of intervention that is only implemented when it serves the specific interests of the interest group parties. Thus slum demolition is a process in urban renewal, a process in urban planning, which has become a mode or tool of the ruling classes for fulfilling their own interests. In shaping the city in such a way that it conforms to the upper class notion of the city, it is its interests, which are furthered. In actuality clearance is necessary to make the centrally located areas available to the capitalists for various economic activities.

As far as Indian urban policies has been concerned it has been seen that there has been a lag between the promises made in different plans in paper and actual practical work that has been done. Though the state within its set structure of the society has tried to work for the poor through different urban planning policies and housing policies like the different slum development policies in Mumbai, urban land (ceiling and regulation) act, more peoples participation in development activities through 74th constitutional amendments, different poverty alleviation programmes but in actual reality very less has been done. The politicians, builders, businessmen, slumlords, the elite class has continuously manipulated the different policies according to there need and profit maximisation motivations and left the urban poor in constant misery. Urban policies in general and urban planning in particular has been become an instrument in the hand of the capitalists to fulfil there needs. Thus there is need to make urban development pro-poor or to evolve an urban development framework to agree to the needs of the most vulnerable sections of the urban population.

They should not be concerned about the needs of the capital alone. This can be done through two things. One is the representation in the planning and policy-making bodies and other is the creation of mechanisms or forums for participation in policy making. Thus decentralization of decisionmaking should be done as has been done through 74th constitutional amendments. But the decentralization should mean change in the structure or power sharing in the society and not decentralization of some convenient functions or responsibilities in a centralized society as of now where economic processes and decision making paradigms are centralized. So decentralization is not only about taking it down to the community but looking at how communities themselves can determine as stakeholders what should happen around them and whatever happens around them should be for them.

People's participation, decentralization, and Privatisation should not be taken as synonyms of each other. Role of the state and role of private sector and people's needs should be appropriately and adequately discussed so that some general understanding is arrived at, as there cannot be a single model applicable to all the urban housing and basic services and utilities. All dimensions, political as well as economic, of the 74th constitutional amendment be analysed and understood realistically and not on ideological or emotional basis. A proper understanding of urban institutions and their functioning has to be developed. The outcomes of NGO's acting as catalysts in the community actions towards development are very encouraging. The experiences and experiments have been regarding alternative institution building, where the stakeholders themselves are directly brought as actors in the development process.

But NGO's as the solution for housing struggle of the urban poor has also been questioned upon. This has been mainly because their internal limitations set to them like the financial constraints and the external limitations like the power politics and administrative constraints. Housing today is looked upon merely in real estate terms. This is what the real –estate agenda has encouraged today due to the Privatisation thrust in housing and corporatisation of the various development and construction activities. Housing projects are evaluated in terms of size, the built up area, the FSI consumed, the financial turnover, and various other business and marketing merits. The bigger the project, the better it is and the greater the attraction for developers in undertaking the scheme. A huge network is thus established between the developers, the landowners, and the financial institutions wherein the slum dwellers find no place.

Thus there is need of changing in the attitude of the government and the elite towards the slum dwellers. Programme for slum development must primarily be seen as an environmental scheme and not merely as an agenda for real estate development and construction turnover. It is the slum-like conditions *i.e.* lack of drinking water, inadequate toilet facilities, garbage, heaps, lack of sewage disposal, absence of open-spaces, inadequate and unsafe access that are of primary concern. Even with the rehabilitation programmes by the Government and the NGO's some basic element is lacking in these projects that is the people. This has been particularly the case of the rehabilitation of the evicted SGNP slum dwellers.

It is obvious from the above argument that neither the state nor the private sector alone can handle the problem of housing for the poor. So there is need for the NGO intervention. But NGO's have there own drawbacks in terms of financial and administrative constraints. So the solution to overcrowding and housing is to ensure that the socioeconomic compulsions that force mass migration towards the cities are ended. However another more rational way out will be the integrated efforts of the private sector, central government, civil society and people. However, this requires simultaneous efforts by national, local governments, civil society, and private sector and people themselves to eliminate impediments at all the levels. While central governments address policy matters and regulatory impediments nationally, local authorities and civil societies should design strategies to make appropriate interventions and regulatory changes in the city. Local experiences should be fed back to national governments to influence their support to cities, as well as for redesigning national programme.

Another thing that can be done is the site and service scheme that has been started by the government in Fourth plan and continued in seventh plan. This is a solution by which the 5 million homeless workers of Mumbai can be housed. In this scheme the government is totally responsible for providing all the services such as water, electricity, sewage, drainage, and toilets, for providing technical know how and skills, interest-free loans and for providing building materials at highly subsidized rates. First of all land is taken over under the Urban Land (Ceiling and Regulation) Act at the rate of Re1/-a square foot which is virtually free or at adequate compensation or if land is already in possession of the government it can be used directly. Instead of This land is divided into thousands of small plots. Then no construction is actually done in the beginning, as is the case of conventional schemes. But water supply, sewage drainage and electricity are provided to each plot. Then, depending on the financial capacity of each person the worker can build his house. The advantage is even a person with a budget of 2000/-can go ahead. Secondly the house can be developed over time. Thirdly and most important, the worker has control over his house. Housing must be made fundamental right enshrined in the constitution. Both the availability of direct finance and subsidized building materials will tremendously excite the slum dwellers in under taking the renovation and reconstruction of their houses on their own. From the table given we can see that people with different income group will be able to afford different types of shelter within there limited finances. Like People with income of as low as Rs.200/-can afford a house of cost 3,200 where the site and services has been provided by the government and the person can make a house within his or her budget. Whereas, a person with a monthly income of more than 1500, can afford a house of Rs.56, 000 made in a conventional way.

Also under the development proposal, the house will be ground and ground + one or two stories high, enabling easy repairs and maintenance directly under the control of the users. There would, therefore, be less dependency on hired skills and services. This will encourage people's participation in decisionmaking and will inculcate a greater sense of belonging resulting in personalization of spaces and structures. While the construction of houses will be the individual's prerogative, the restructuring of the slum-layout, road, services, open spaces etc., could be a collective effort with governmental support

Development programme for each slum will have to be evolved independently and relevant guidelines fixed. Even F.S.I for each slum may vary to enable housing for all the slum-dwellers on, as–is-where–is basis. Each slum has its own peculiar situation and needs. For example what will apply for Dharavi may not be relevant for another slum in Jogeshwari and vice versa. Therefore, within the main policy framework, individual development strategies will have to be evolved. This will encourage people's participation in decisionmaking.

PROMOTE SUSTAINABLE DEVELOPMENT IN FRINGE AREAS

Control of Development on Fringe Areas: In metropolitan cities and mega cities, urban development is mostly in new settlement areas and new activity centres with planned infrastructures and facilities in the fringe areas to accommodate the increasing population and activities. Unplanned urban sprawl grows around such centres on the agricultural lands taking advantage of the nearby facilities and infrastructure. In such cases, from an environmental perspective, regulations for protecting the agricultural and vacant lands by restricting developments and the stipulations of the regulations may be as below.

i) No use, other than agriculture or irrigation facilities, is permitted.
ii) Existing water bodies to be preserved.
iii) No new building or extension of any existing building exceeding the height of 3.75 metres shall be allowed subject to the total covered area of 50 sq. m.
iv) The minimum front and side open spaces shall be 2 metres and the minimum rear open space shall be 5.00 metres.

Redevelopment of Blighted Pockets in the Central Area: Blighted pockets of low key commercial areas are observed within the core areas which, due to the small sizes and multiple land ownerships, do not get redeveloped as per the existing regulations. If these areas can be redeveloped in a planned manner much of the demand for commercial floor spaces can be served. For the redevelopment of such areas incentives towards land assembly by amalgamation of plots and additional floorspace ratios need to be given. The regulations in this respect may be:-

1. For land assembly exceeding one acre in size, the additional floorspace ratio would be 20% above the permissible limit.
2. In case of land assembly of less than one acre the additional floorspace would be 10% above the permissible limit.
3. In such cases the developments should confirm the zoning regulations.

Compulsory Rainwater Harvesting in New Area Development: Private developments in the form of sub-division of mother plots (for plotted developments and for apartments) is taking place in the fringe areas and adjoining municipal and non-municipal areas. There are regulations in most of the plans which vary according to the size of the mother plots and the regulations specify the minimum width of roads, the percentage of open spaces, the land for physical and social infrastructures *viz.* drainage, water supply (pump house & water treatment plants), sewerage (sewage treatment plants or oxidation ponds), school, health centre, market, milk booth, post office, power substation etc. In order to comply with the present efforts towards utilization of natural resources by rainwater harvesting and ground water recharging, the regulations for subdivision should include the mandatory provision of community pools of sufficient size so that the rain water from the area can be stored in such pools. The water may be supplied to the community for uses such as gardening, car washing etc. Adapting physical planning to promote sustainable development: efficient infrastructure planning.

Prescription of street alignments for regional roads in fringe areas : The regional roads (National Highways, State Highways or District Roads), connecting a city with the hinterland are often constricted in the fringe areas due to lack of scope for widening owing to dense developments and abutting built up areas. To avoid such situations where the regional roads pass through vacant areas in the fringe of the city, advance actions for prescription of street alignments may be made without acquiring land. The proposed right-ofway as per future requirements may be prescribed and the regulations may be as below.

i) For any development on the adjoining plots on the regional roads, the owner/developer would have to make a setback following the proposed right-of-way line and the owners/developers would be allowed the same floorspace as they were eligible for the original plot.
ii) In such cases after the alignment is notified no subdivision of the adjoining plots would be allowed.
iii) In cases where the plot size is such that no development is possible by allowing the set back, the local body would have to acquire it.

Dispersal facilities around transport nodes: The railway stations, the regional bus terminals/stops are the important transport nodal points of the urban areas. These areas get congested with dense unplanned commercial and residential developments and in course of time become too congested for easy dispersal of passenger and vehicular traffic. Given the urban growth rates future public transport will carry larger volumes of passengers. Therefore, the areas required for dispersal facilities will need to be increased. Where such areas are already congested, redevelopment plans are to be prepared by the local body and the redevelopment actions may be initiated, including through private-public partnership projects. In less pressured situations regulations for control of development should be made at least for the area within 200 metres on all sides. In such areas provision should be made for adequate parking facilities for different categories of vehicles including the parking and loading/unloading facilities for transit and paratransit vehicles.

Adequate width of the connecting roads, exclusive roads for pedestrians or grade separated pedestrian facilities should be made. In order to achieve these, specific plans should be prepared indicating the future right-of way of the roads, the parking areas for different categories of vehicles, the pedestrian-only roads and the integration with the railway station or the bus terminal. The regulations in such areas may include the following:

i) The Floor Area Ratio (FAR) for buildings of different use categories and means of access would be half of the permissible FAR in other areas.
ii) The compulsory provision of parking spaces for buildings of different use categories would be double of that required in other zones.
iii) No new cinema halls, theatres and entertainment centres would be allowed within the area.
iv) For buildings with retail commercial in the ground floor, the minimum front open space would be 5 metres.

Environment protection around relocated hazardous uses: For implementing the development plans, the non-conforming uses located in a scattered way are required to be relocated in the specified areas in the fringe within a stipulated period. Such uses may be the obnoxious and hazardous industries, tanneries etc. Normally, such areas for relocation are selected beyond the city limits and in the vacant and agricultural lands.

When such relocation of activities takes place, unplanned developments adjoining such centres occur and gradually these areas grow.

To avoid further environmental hazards, the relocation areas should be provided with a buffer zone where no developments other than agriculture and pisciculture are permitted. In this regard, regulations of the Pollution Control Board are to be followed. But in the development plans there should be specific regulations in respect of the buffer zones. The regulations in this respect may be:-

i) The area covered by 200 meters on all sides from the boundary of such areas would be designated as Buffer Zone.
ii) No developments other than agriculture, pisciculture and plantation of trees would be allowed within this zone.
iii) The existing residential uses within the Buffer Zone would have to be relocated within a stipulated period.

Adapting physical planning to promote sustainable development: conservation of water resources and waste management

Protection of Water Fronts: The water fronts (sides of rivers, canals, lakes and big ponds) in many cities are encroached by unauthorized users and developed in an unplanned manner. These water fronts need to be protected to ensure proper drainage, and access for open-air recreation, water transportation and protection against soil erosion. Area within 100 metres from the banks should be designated as Water Front zone and specific regulations should be prescribed. The regulations for such zone may include:-

i) No new building within 30 meters from the edge of the banks would be allowed.
ii) In the area lying between 30 metres and 100 metres from the edge of the banks no building more than 5.00 metres in height and 30 metres along the waterfront would be allowed. In case of buildings on stilts the maximum height of the buildings shall be 6.50 metres.
iii) There shall be a linear gap of 50 metres between two buildings alongside the water front.

Environment protection around solid waste disposal sites: Solid waste management is one of the most critical problems of cities.

The locations of the intermediate collection sites and the final disposal grounds need special attention in consideration of the environment hazards of the nearby localities. The intermediate collection sites are generally located within or near the settlements and therefore need to have a buffer zone. This buffer zone should cover at least 30 metres on all sides. The regulations in this respect may be:-

1. The intermediate collection site may be designated as Inner Disposal Zone.
2. The area should be provided with boundary walls of at least 3 metres high on three sides.
3. The actual dumping area should be circumscribed by two to three rows of trees in the buffer zone.

LAND AND BIOTIC RESOURCES

LAND-USE CHOICES

The small size of many island developing States makes competition for land more severe than in larger countries. Large commercial plantations were frequently established under the colonial system, and continue today in the more fertile plains of island countries, forcing many small farmers to resort to the poorer soils in hilly regions. This, combined with poor land use practices, has resulted in deforestation and soil erosion.

Due to the peculiar geological characteristics of many islands, water resources are scarce, often limited to thin lenses of freshwater floating on sea water, recharged by rainfall. Poor management of watershed catchment areas and of water flowing through agro-ecosystems, particularly irrigation supplies, can lead to water scarcity and pollution of downstream supplies. Population pressure, expanding tourism, and reduced precipitations of recent decades, have resulted in the over-exploitation of groundwater and seawater intrusion, with subsequent further deterioration of water quality and quantity.

Land use planning is a prerequisite for long-term sustainability, since without it, the economic and social fabric of society as a whole will be endangered. Advice on area-specific soil and water conservation, and management practices will be needed for farmers, graziers and others. Such advice will only be adopted if it is perceived as being capable of bringing rapid economic returns to the users without threatening their fragile survival strategies. Success will also depend on guaranteeing long-term rights to land and water resources.

"Land Use Planning, Management and Information System in Grenada"

Grenada is highly dependent on agriculture, deriving 90 percent of its export earnings from this source, though the sector has stagnated over the past three years due to a combination of low export prices and declining productivity of the traditional export crops. The agricultural development strategy of the Government is to improve the productivity of traditional export crops, while diversifying into nontraditional crops for export and for agro-industrial integration into the expanding tourist sector. Emphasis will also be placed on increased domestic food production. The strategy involves bringing idle private and public lands back into production, and strengthening private and public sector institutions. Topography is a key factor in implementing the above strategy.

On the islands of Grenada and Carriacou, approximately 77 percent and 54 percent respectively of the land area has slopes exceeding 20 degrees. Only 33 percent of the soils are deep, well drained, and free of cultivation constraints. The rest, apart from excessive slopes, are shallow or rocky, or have seasonal drainage problems. A favourable feature of Grenadine geography is a well-defined system of watersheds governed by the topographic position of the rivers and streams.

The system of watersheds offers the opportunity for the establishment of rational land use consistent with the characteristics and conditions of each watershed, and with the need for soil and water conservation. The main objectives of this newly launched project (1993-94) are to create the technical basis to support rational development and conservation of the country's natural resources, and to develop an effective forestry management policy, and forest area conservation and development programme for the country.

The project will provide specialised technical assistance and practical hands-on training to local technical staff in the establishment and development of computerized natural resources information systems. The project is designed to provide a rapid response to an immediate need affecting the country's food production and the degradation of its natural resources. It is expected to contribute to establishing a sound basis for natural resource planning and policy formulation at both national and watershed levels. More specifically, the following needs to be emphasised:

- sound cropping practices and cropping systems which maintain soil fertility and productivity for sustainable crop yields as well as practices such as soil testing and analysis to determine fertilizer recommendations on a soil-and crop-specific basis. Land evaluation/ land use planning are pre-requisites for sound land use and crop diversification. Development of land information systems will make a vital contribution to improved resource management;
- suitable water harvesting methods, both for domestic and agricultural use (supplementary irrigation), aiming at significantly increasing the groundwater recharge as well as collection and storage of rainwater from ground catchments;
- use of biomass residues, as a source of plant nutrients, will reduce use of mineral fertilizers in favour of nutrient sources available within the farm. This would contribute to environmental protection and decrease pollution of soils and groundwater.

Land conditions are similar in all coral island countries, and, apart from those affected by human action, soils are homogeneous. The main restrictions on land use are urban encroachment, tourism, established tree crops, especially coconut, or the existence of improved soils devoted to market gardening. Where choices for land use are still available in these islands, integrated coastal area management should be applied in a comprehensive programme of land use planning.

In small low-lying coastal areas, land well suited for agriculture is under severe pressure from expansion of housing, industry, and tourism. Under the market system, with private land ownership, the immediate economic returns from such uses of land always override the longer-term benefits to be derived from sustainable but low-return agriculture.

Economic pressures thus make it difficult for the rural sector to opt for the sustainable use of the resources entrusted to its care, such as:

- light use of coastal mangrove forests, while maintaining the additional functions of nursery areas for marine species, and protection against major damage to adjacent land from cyclones or tidal waves;

- brackish-water fish ponds in humid climates or salt pans in and or semi-arid conditions;
- cultivation of wetland rice or appropriate dryland crops, irrigated or rainfed, with flood protection and water control (mainly drainage facilities);
- cultivation of annual crops, adaptable to changing conditions, or perennial crops, entailing long-term decisions and larger investment; and
- conservation of local germplasm offered by traditional farming systems.

Hilly or mountainous areas face a different, but no less difficult, set of choices. Depending on climate, geology, landform and soils, the land may have high or low potential for nature conservation, forestry, grazing, production of perennial or long-term crops or cultivation of annual crops. Conditions can vary greatly over short distances, and the optimal response to a country's needs and objectives may well be a mosaic of different land uses and protected areas.

In small island developing States, the relationship between forests and trees, and other types of land use and economic activities, is critical. Forestry should be systematically represented in national and district planning commissions, land use zoning activities, town planning, energy and infrastructure strategies, as well as in planning for generation of income, employment, and food security. Forests and trees and the products and services they provide must be validated and valorized in view of their role in the protection of water resources and of the environment in general.

"Tropical Forests Action Programme"

Forest resources contribute to environmental protection and food security. Rural people exercise important customary rights over forest products, which have traditionally been used sustainably. Current forestry problems originate, among others, from official policies which are not supportive of sustainable management and utilization of forest and tree resources, The importance of forest resources has motivated many Caribbean island countries to develop appropriate inter-island forestry and environmental networks to promote forestry research, as well as carrying out Tropical Forests Action Programmes (TFAP) in order to implement the sustainable management of forest resources. The TFAF planning phase in Antigua and Barbuda, Barbados, Dominica, Grenada, Montserrat, St. Kitts and Nevis, St. Lucia, St. Vincent and the Grenadines, and Trinidad and Tobago is concluded and the implementation started in November 1993. This multinational exercise defined regional issues and produced key documents, including a survey of forest management and silvicultural systems in the region.

In the Pacific, Fiji is already implementing a TFAP-framed national forestry action plan, Vanuatu and Solomon Islands are starting the planning phase, and others (Kiribati, Micronesia, Tonga, Tuvalu, and Western Samoa) have expressed interest in launching a TFAP process. A TFAP Training Workshop for this region is scheduled for May 1994. In Africa, Cape Verde Islands are in active preparation of their National Forestry Action Plan, and a recently held national Forestry Congress has highlighted the overall progress achieved and original aspects of forestry in the country. Legislation and planning exercises are under way in the Comoros Islands.

WATERSHED MANAGEMENT

Urgent implementation of watershed management and rehabilitation programmes in most small island countries is needed for sustainable development. On most small islands, distances between highlands and coastal areas tend to be short. Under such conditions, the role of forest ecosystems as regulators of water supplies for the island crop environment and soil formation and protection is especially critical. By extension, the sustainability of agricultural productivity and potential greatly depends on the regulatory functions of forests.

Protecting forests and trees is essential for warding off environmental degradation and rural poverty. The multiple goods and services available from forests offer many possibilities for off-farm employment and for economic diversification, with low capital input, for example through the integration of plantations with agricultural crops, in soil conservation and watershed management, as recreational areas, and as a source of subsistence.

In spite of the significance of forests and tree-based resources, present trends are not encouraging; forest resources continue to be poorly managed and not used rationally in many island countries. Deforestation occurs due to clearing for agriculture in areas not suitable for this purpose, culling for fuelwood, and urban and infrastructure development. Population growth and economic reforms pose new and increasing demands on the forestry sector. The result is accelerated loss of forest cover and of its associated benefits.

"Watershed Management and Conservation Education in Western Samoa"

This UNDP/FAO project (1992-95) looks at the inter-relation between the management of upstream and downstream areas within a given watershed. This has been shown through the implementation of soil conservation and watershed management measures such as conservation plantation, agroforestry practices, plantation of forest and fruit trees by farmers, conservation farming systems, extension and education programmes for different target groups.

A monitoring system has also been developed to increase understanding of the impact of upstream management on water quantity and quality and other downstream values.

The destruction of vegetative cover by frequent tropical cyclones has led to accelerated soil erosion and landslides, with a direct impact on the watersheds, agricultural production, water supplies, hydropower generation, lagoons and reefs. The rehabilitation of upstream areas and the participation of the local population in conservation of downstream areas an integral parts of the project. This will facilitate the conservation of flora and fauna and improve the socio-economic condition: of watershed dwellers.

GREENING ACTIVITIES

Considering the limited land area and the islands' comparative fragility, the role of vegetation takes on particular importance. Some islands have no land suitable for agricultural activities and are confined to producing vegetables and growing fruit trees. The cover of low-lying islands is composed of mangroves and other plant associations, as well as exotic species. Demands for boats, construction material, rural housing, resort development and fuelwood have resulted in depletion of tree cover. This has affected the dynamic influence of ocean, coral reefs, land formations, and vegetation on each other.

Drought and desertification in some small island developing States renders afforestation difficult. In order not to interfere with endemic species and corresponding life systems by introducing exotic plants, it is preferable to use endemic species (bringing them in from nearby islands if local vegetation has been degraded) when undertaking forestry plantation and afforestation. Current water harvesting techniques, including soil manipulation devices aimed at concentrating rain water to better benefit planted seedlings, make plantation efforts in dry environments less uncertain than hitherto.

Further research is necessary for:

* improved knowledge of the original indigenous plant species and formations;
* selection and use of local species suitable for incorporation in soil and water conservation, range regeneration and fodder production, and fuelwood supply; and
* screening and eventual introduction of suitable non-indigenous species tried under similar conditions for use in all the activities.

Greening activities should also include the identification, protection and proper management of natural formations.

"Development of Forest Resources in Cape Verde"

Within the national development framework, this project (1984-93) offered institutional support to the Government for the establishment and consolidation of an efficient forestry administration and the creation and management of enhanced forests and rangelands to promote rural development. These activities will contribute to combating, desertification, rehabilitate the environment through the reconstitution and conservation of the vegetative cover, establish integrated production system, contribute to the provision of wood for multiple uses (energy, construction, etc.), enhance socioeconomic welfare through the sustainable exploitation of forest resources, and transfer management responsibilities to a competent administration working within an adequate institutional framework. This project, completed in December 1993, should 'facilitate, *inter alia,* the finalization of the National Forest Action Plan, leading to better coordination in the planning and implementation of sector policies while promoting the participation of rural people in the preparation and management of the action plan.

PROTECTED AREAS AND BIOLOGICAL DIVERSITY

Island bio-geography has led to a high endemism in terrestrial species, especially in larger islands. All small island developing States are particularly vulnerable to the impact of introduced species and deleterious human action.

The physical characteristics, the biotic formations, and socio-economic situation of small island developing States determine the conditions and role of forestry in sustainable development. The implications of their small size, their isolation, and the nature of their forest resources define the available models and opportunities for forest development, and suggest the necessary precautions to be taken in any manipulation of plant and animal populations. The increasingly negative impacts on local economies and on the quality of the environment are already visible in many cases through shortage of water supply, soil erosion, increased sedimentation of rivers, estuaries and coastal areas, and coastal erosion. Deforestation and forest degradation have led to the extinction of many animal and plant species, causing irreversible loss of genetic resources and ecosystems. The establishment of protected areas, such as forest reserves, national parks and wildlife sanctuaries, supported by botanical gardens, herbaria, etc., are essential for conservation of biological diversity in these countries. Those areas can become the basis for eco-tourism.

Marine ecosystems are prone to destruction of habitats and bio-diversity. There are several causes of this, including: destruction of coral reefs (both deliberately by fishermen or tourists and indirectly by pollution, sedimentation, and land reclamation for building sites); conversion of mangroves and wetlands (resulting in loss of important nursery areas); the use of long driftnets (mostly impacting marine mammals, turtles, and non-targeted fish) by a growing number of foreign fishing vessels; and more generally over-fishing. Lack of monitoring makes it difficult to quantify the damage of such activities on marine life. Surveys are needed to identify representative habitats and ecosystems for the establishment of marine reserves and parks.

4

Forest Land and Forest Rights

Dediapada is the most backward Tehsil (block) of Gujarat in terms of income, physical amenities like road, electricity, drinking water and other development parameters. The only good thing was a good forest cover, mainly of teak and bamboo, one of the last few remaining forests in Gujarat. In the name of protecting and preserving the natural resources, the area was declared 'Shoolpaneshwar Wild Life Sanctuary' in 1987, which would take away whatever meager economic opportunities and rights the tribals had. When we started getting involved in the problems, we had no idea that it would turn out to be a more than 20 year long struggle: from preventing their eviction from the forest land to getting property rights on these long-occupied forest lands and forest resources for the tribal families, from saving the teak and bamboo forests of the sanctuary area, which were being plundered by the state itself, to getting for the tribal communities management rights over forests. One problem led to other deeper and more intricate problems with unknown ramifications. People were subjected to all sorts of harassment by the Forest Department (FD) like beatings, bribery and eviction, because the lands they were cultivating for years were the property of the FD and the people had no rights.

First time in the area, tribal men and women filed criminal complaints against the erring forest officials. We filed many petitions in the Gujarat High Court against the eviction drives by the FD of the tribal families from their long-occupied lands. These pieces of lands were the only means of their livelihood. The stay orders by the High Court gave a foothold to the tribals for sustenance. Then ensued a long struggle of tribal organization to get property rights on their long-occupied lands.

It was becoming very clear that the FD, which was bent upon to evict the tribal families from the small pieces of lands, was neither interested nor able to protect the few remaining forest resources also, leave alone the wild life. After the declaration of the Sanctuary, it continued allowing a powerful paper mill to exploit and take away thousands of tons of bamboo from the core sanctuary area every year on a throw-away price of Rs. 100 per ton of bamboo, which was previously Rs. 5 per ton! Two faces of the State were too visible! People opposed the bamboo cutting at the risk of losing labour wages for the bamboo cutting.

Fearing that the bamboo cutting activity would deplete the whole area and leave them without an important support resource, the tribal villagers decided to oppose the cutting. Then ensued a long struggle of the people and us against the paper mill, forest department, local politicians, contractors and other powerful vested interests, to save their precious bamboo. They made several representations to the authority, but in vein. They were compelled to stop cutting work and prevent trucks plying to carry out bamboo. Their main demand was that they should be given rights on the bamboo, which they would regenerate and sell to the paper mill. That way the resources would be properly managed, protected and would give a good source of income to them, who were otherwise totally cut-off from the mainline business and industries and development.

But the state would not budge an inch. Then the Gujarat High Court intervened. And the situation became even murkier. Ignoring their own commission report stating categorically that Shoolpaneshwar Sanctuary was a sanctuary of cut bamboo and also the basic legal points, the division bench consisting of the then Chief Justice of Gujarat High Court, on a pretext of a non-existent permission letter by the government, granted the cutting. Bright red seal of the Gujarat High Court smoothened the way to take away thousands of tons of bamboo from the sanctuary! People were reduced to helpless and hapless bystanders to the loot of their natural resources. What they could save with the force of their collective resolve, was being driven away by a fiat of the highest Court of the state. Next time, the Central Bureau of Investigation (CBI) ordered by another judge of the High Court, confirmed our reporting and held forest officials and mill personnel guilty of several offences under various forest-related acts and on the corruption charges. Again the Court dismissed the CBI report on technical grounds. Bamboo cutting continued. It had now the legitimacy of the judiciary.

The story of a series of violations of the existing laws and regulations and their cover up is deeply buried in the piles of affidavits and reports filed by the State and the mill company and ARCH's detailed rebuttals of their claims and revelations. The Court never asked the State to respond to their irregularities and unlawful activities. A curtain was dropped by the Court on the whole sordid story of decimation of bamboo in the heart of the Shoolpaneshwar Sanctuary. The Central Empowered Committee (CEC) appointed by the Supreme Court also did not pay heed to our contempt plea that the Higher officials of the Forest Department had violated even the SC's order of 14 February 2002 order of not taking out anything, even dead, dried leaves also, from any Protected Areas of the country and on the face of such clear order had even extended the lease given to the paper mill of cutting thousands of tones of bamboo for 10 years! Of course, the CEC issued a stay order. People were witness to the series of violations and illegalities and also the decimation of the lush green bamboo, to which everybody-the Government of Gujarat and FD, the mill, and even the highest judiciary of the State-contributed their might.

Many facets of the state, either in the form of a Welfare state, or in the form of an architect of the 'Modern Temples', or as wearing a mask of custodian of natural resources stared us in the face. We realized that government taking upon itself all the responsibilities and the power to fashion the poor people's lives to pull them out of the poverty-trap and to bring forth better life for all concerned looked a very attractive and seductive dream, when we had started our journey years ago. The actual experience taught us what a nightmare it is! We dwelled deep into the fundamental issues of poverty, development, environmental degradation, laws, traditions, initiative, capacity and potentials of poor individuals.

We realized that the State, which had acquired tremendous powers in the name of protecting the tribal interests, was in fact causing greatest harm and injury to them. Non-tribals are prohibited by law from purchasing any land or property in the tribal areas in name of tribal protection. But at the same time, the State can compulsorily acquire any land or property for 'Public Purpose', which has been too broadly defined. And most importantly, it has completely alienated them from the most important natural resources of their habitat – forests and minerals – by establishing full ownership and control over them. The imperial British government initiated this process by enacting the first Indian Forest Act in 1878 and the independent India further consolidated the process.

That this alienation of the local tribal communities from the forests was responsible for their abject poverty was obvious to us. But that it was not good for the forests too was initially not so apparent. With all the power and authority over natural resources, the state has failed to protect them. Forest cover in the Narmada valley once famous for its teak, bamboo and Mahuda trees had been denuded in the 1970s by the FD itself. And we witnessed the state's plundering of the precious teak and bamboo from the very heart of the sanctuary. We were gradually getting perturbed by the state of forest resources. The depleted forests, denuded hills, degraded land and massive soil erosion were everywhere to be seen. And at the same time, people living in and around the depleted forest area and surrounded by denuded hills have no development opportunities, which they eagerly want. Are these issues really that incompatible as they have been made out? All the years, we were wondering, fuming, struggling to find out a solution to such a bizarre situation, where land is abundant and there is no dearth of human resources, yet no forests worth the name exist. What are the missing ingredients? We realized that it is the government ownership of natural resources and the absence of markets, and not the markets, that encourage environmental degradation and over exploitation and abuse of natural resources. We could also see the outlines of an alternative way for the management of environmental problems by establishing a system of well-defined property rights to natural resources. As this approach involves transfer of rights to resources from the government to the local communities, which are often poor, it also has the potential of giving direct economic gains to the poor while halting and reversing the environmental degradation and neatly takes care of the false dichotomy of Development vs. Environment.

IMPACT OF TOURISM AND TRIBAL RIGHTS

Tourism has fetched some benefits to tribal areas, but its overall impact has been negative. The limestone caves with formations of stalactites and stalagmites, located in Borra panchayat of Visakhapatnam district, are among the world's largest. Dating back to the Paleolithic Age, the caves, spread over two square kilometres, have been sacred to the tribal people of the region ever since a shepherd chanced upon a stalagmite in the shape of a Siva 'linga' deep inside the caves several years ago. Now, with the Andhra Pradesh government promoting tourism in the area, the caves themselves have become a major livelihood issue for the tribal people.

Since their discovery, the caves have been preserved and nurtured by the tribal people of Panasaputtu village in the panchayat. They worship the 'linga' during Mahasivarathri. The other naturally sculpted splendours in the caves and the three cascading waterfalls deep within make them a major tourist attraction. Also, the tribal people earn some money acting as guides. In 1977, stairways were cut into the cave walls and the tribal people used kerosene lamps and torches to guide tourists. In 1992-93, the Tourism Department took over the Borra Caves and, started to levy an entrance fee. The tribal guides petitioned the government and managed to get a 25 per cent share in the collections, invoking the "social responsibility" of the state. From the entrance fee of Rs.10 a person, Rs.2.50 was given to the guides and the rest went towards maintaining the caves. According to the 20-member Guide Union president S. Indore, on an average each of them used to get Rs.7,000 a month.

In 2002, the government, in order to "beautify and maintain" the caves, cemented the floor, lit up the interiors with 74 halogen lamps that burn throughout the day, and stopped giving the tribal people their share of the fee collections as it wished to "get back the money invested". The entrance fee was raised to Rs.25, and a fee of Rs.10 was collected from tourist using a still camera and Rs.100 for a video camera. The caves were also "rented out" for film shooting at Rs.10,000 a day. Twenty tribal people were "appointed" as guides on a contract, which expires this year, by the Tourism Department on a monthly salary of Rs.3,000, that is, at the rate of Rs.100 a day. As per the contract, the guides have to work from 9 a.m. to 9 p.m. and even a day's absence would entail a Rs.100 cut from the salary. The tribal people had no choice but to accept this offer although it meant a substantial drop in their income. According to Ravi Rebbapragada, executive director of the non-governmental organisation Samata, the Tourism Department gets Rs.25 lakhs annually from entry fees. Had it continued the earlier arrangement of giving the guides 25 per cent of the entrance fee collected, they would have got an annual income of Rs.6.25 lakhs.

Also, they would have been stakeholders and would not have felt alienated from the development of the area. According to guide J. Pandu, during peak season (April to December) over 1,000 tourists visit the cave every day and some 150 during the lean season (January to March). Between mid-November and mid-December, during the Karthigai festival season, the numbers touch 2,000 a day. According to guide B.B. Ramachander, on some days duirng the peak season the collection touches Rs.1 lakh. Says guide P. Anand: "We have made representations to the Tourism Department about giving us a share in the earnings. Our panchayat has decided to launch a struggle if they do not relent by the year-end, when our contract ends." Says Indore: "Nothing works without a struggle."

The income from tourism seems particularly important to the tribal people, as their income from farming has fallen owing to intense film shooting in the area. According to K. Venkat Rao, secretary of Tribal Education Rural Development Society (Terd), one of the 10 community-based organisations under Samata, several popular films in Telugu and Tamil were shot in these locales. So much so that film shooting itself has become a major tourist attraction. In fact, the government did not want to dismantle the sets of the popular films hoping to attract tourists.

The Tourism Department wanted to convert a 'botanical garden' on 250 acres into a permanent structure but representations and dharnas forced it to give up the idea. According to T. Hari of Panasaputtu village, film-shooting "has spoilt our land". In order to put up massive structures the topsoil is heavily disturbed. This prevents seepage of water and affects the water level. Disturbance of the topsoil leads to a substantial fall in productivity. According to G. Neelakandan of the village, this is a typical case where the local people are not only alienated from development but are a victim of it.

The tribal people are alienated from land directly as well, with the government refusing them land pattas. The 14 villages in Borra panchayat have always been in the eye of the storm. To begin with, as they are located in a reserve forest, there has always been an ambiguity about the land-whether they are revenue land or forests. The State government denied the tribal people land ownership saying they were forests, although it gave out mining leases to private companies. After the Supreme Court cancelled the mining leases in 1997, the State government turned to tourism, which is also alienating the people from their land and livelihood.

Under the Forest Conservation Act, 1980, when land was divided between the Forest and Revenue Departments in Borra panchayat, 1,150 acres (460 hectares) came under the former and 426 acres were left out of the reserve forest area to be distributed to the tribal people. But overlooking the Land Transfer Regulation Act, no patta was given to the tribal people of Borra panchayat.

In 1998, another NGO, Adivasi Mitra, along with Terd filed a petition in the Andhra Pradesh High Court. In 2003, the State government relented, and gave the tribal people temporary, or D-Form, pattas, which gave the government the option of taking back the land any time without giving any compensation. One such case is Punnasaputtu village, where all the tribal families have been given only temporary pattas for the 76 acres (30.2 hectares) they have been cultivating for generations.

Although their farming is rain-dependent, they managed at least two crops a year of paddy, pulses and millets. Their orchards produced mangoes, jackfruits and gooseberry. The tribal people have been protesting but are scared to take the struggle further as they do not hold permanent pattas. Says Hari: "We have been cheated by the government. Our lives are deteriorating with development. But without title deeds to our land we are helpless."

Sub-Collector Ch. Appalaswamy, reacting to the fears of the people, says: "We have made an amendment to the D-Form. Now we do provide them a compensation when we ask them to move, provided, of course, they have put up some structure on the land." But what of the farmlands where there are no structures? And, where will the tribal people go with the Agency area already occupied? Appalaswamy has no answers.

FOREST RIGHTS ACT TRAMPLES TRIBAL RIGHTS

Over two years after the FRA came into force, a visit to Jharkhand and Chhattisgarh – two of the Naxal-affected states – showed that instead of getting the promised benefits, tribals and forest dwellers are, on the contrary, engaged in a desperate fight to save their home, hearth and source of livelihood. A large number of them face displacement as the land they have lived on for generations is in danger of being handed over to various industrial and mining companies. The FRA, in effect, has remained on paper. It has conferred much too little to far too few: a maximum of about half an acre to less than one per cent of the families covered by the Act. As for village community rights over pastures and forest produce, these are yet to be taken into account. So far, a total of two persons have "benefited" in Latehar district of Palamau division from the Forest Rights Act. These two adivasis from Durup village were allotted between 0.2-0.4 acres of land – barely enough for sustenance. The actual, average size of land villagers have is between 1.5-4 acres. Villagers all over Palamau division, waiting for completion of the process of recognition of their rights over lands they occupy, had similar tales to narrate: "Forest officials tell us we will be given "20 decimal – 50 decimal" (0.20 acres – 0.50 acres) at the most. Those who have five acres will get a maximum of 0.5 acres."

When Palamau deputy commissioner (DC) Amitabh Shukla announced the settlement of claims of around 170 villagers on April 14, many of them were those processed by 'thekedars' (contractors), according to Awadh Singh. The sizes of plots allotted, "averaging one acre" according to the DC, was between 0.4-1 acre, said Singh.

The actual sizes were three to four times this size but the villagers, in awe of 'sahib log', government servants, were browbeaten into agreeing to this. ((Forest officials are seen as the most serious hindrances to full implementation of the law. According to activists, forest officials not only do not extend any help to those wanting to exercise their rights over forest land as stipulated in the law, they resort to various devices to deny them those rights.

The forest official, who is a mandatory member of the joint verification team often does not turn up for meetings to settle claims. A forest department functionary in chief conservator's office said the claims were not being processed fast enough as no claims were being received. "We can clear the applications only when we get them," he said. ((Corruption, too, has muddled the process. According to Awadh Singh, involved in helping villagers in the implementation of the Forest Rights Act, the 'amin' who measures the piece of land claimed by a family, often demands Rs 1500-Rs 2500 from every villager for doing his job. Hence, so far, according to latest data, Jharkhand has settled claims of slightly over 9,000 tribal families. Compare – or contrast – this with the eagerness with which the government has rushed through MoUs (Memorandum of Understanding) with corporate entities, awarding them nearly two lakh acres of land, much of it forest, with utter disregard to environmental and social concerns. In Keradari Block in Hazaribagh district, 1,300 acres of the total about 1,400 acres that have been given to a corporate house is forest land, says leading activist Dayamani Barla. Not a single villager in the area has yet been given any 'patta' or piece of land under the Forest Rights Act.

It is the government and the corporate houses that are aware of the land and resources available; the people who have been given the right over their land are not. Many, even within 30 km radius of the state capital Ranchi, have not heard of the Act. A few who have are waiting for the administration's response, some for over a year, after filing their application forms. They are routinely and regularly called for meetings 30 km away, meetings that do not happen because one or the other government functionary is unable to attend. The FRA and the people it is meant for figure low in the government's priority. In practice, the net effect of government's approach would, rather than secure the tribals and forest dwellers in their habitations, uproot and displace a large section of these people without any meaningful resettlement or rehabilitation because that policy is still not in place.

The government has been on a MoU signing spree ever since the mineral-rich Jharkhand (2000) comprising a third of India's entire mineral wealth came into existence. Indian Social Action Forum estimates that over 1,98,362 acres of land is on offer in 44 projects of various companies with whom MoUs have been signed. There are over a 100 MoUs that have been signed. According to human rights activist Gladson Dungdung, various governments in the state till now have signed 102 MoUs for establishing steel factories, power plants and mining industries, which require approximately 2,00,000 acres of land, which directly means the displacement of approximately 1 million people.

A government that drags its feet in taking action to carry out its obligations under the Forest Rights Act to give tribals and forest dwellers their right to land and forests, has gone into an overdrive to cleanse the forests of real and imagined Maoists through 'Operation Green-Hunt' to make them safe for the corporate houses to move in. The operation happens to be on particularly in areas where the government has promised land to industrialists. There were hardly any signs of the "green hunt" in Palamau-Latehar region, a Maoist stronghold, which has seen trains being hijacked by the Maoists.

FIGHT FOR TRIBALS' RIGHTS

The administration in East Godavari district of Andhra Pradesh has an extremely narrow and short-sighted view on the tribal question: with its characteristic wisdom it has come to the conclusion that it is only CPI(ML) Liberation which instigates the 'innocent' tribals to grab the 'forest' land. Hence, they unleash their rabid police force against the leaders of our Party in this district for ruthless repression. True, our Party is leading the tribals in their struggle for land all along the tribal hamlets of Yeleswaram, Pithapuram, Kotanandur and Tuni mandals braving all odds. But the near-insurgent situation among tribals of Andhra Pradesh has a broader basis and can hardly be suppressed in the long run by dubbing it as mere handiwork of Naxalites. For instance, in the neighbouring West Godavari district also periodical but spontaneous clashes break out between tribals and non-tribal vested interests. To cite one example, sometime in mid-1995, Kondareddy and Khoya Dora tribals in Jillelugudam village offered stiff resistance against goons backed by both Congress(I) and TDP and prevented them from taking over about 1400 acres of cultivable land allotted to them in Jeelugumilli mandal. In this village itself, which is part of a notified tribal mandal where non-tribals cannot acquire land, one non-tribal landlord owns 60 acres of land, all grabbed from the tribals. The tribals are losing their land at an alarming rate. In the notified areas of this district, non-tribals hold 28,367 hectares while tribals own only 13,358 hectares. Losing their traditional land, the tribals are forced to move out of their traditional habitat to eke out an existence as day-labourers in distant cities. According to the 1991 census, the tribal population in 102 villages of the notified Scheduled Tribe mandals of Polavaram, Buttaigudam and

Jillelugudem had come down by 43%. The out-migration of tribals and in-migration of non-tribals into the tribal areas is of such magnitude that today tribals form less than 20% of the population in Jillugumilli mandal.

Our own field studies indicate that it is mostly the moneylenders and the traders who grab away the tribals' land. And the rich farmers from the plains soon follow. Mango traders and bamboo contractors who supply for the paper mills corner thousands of acres from the tribals to raise orchards or bamboo groves. There is an emerging nexus between outside capital and a thin layer of tribal elite and powerbrokers, often belonging not only to Congress and TDP but also to CPI. Not just the local landlord who is invariably a non-tribal, and not only the mango exporters, timber and tendu leaf contractors and paper mills but the World Bank is also a villain which tears away the tribal from the tiny patch of land where he carries on his subsistence farming.

Impelled by the growing pressures from within the western societies against depletion of forests and against newsprint and plywood industries, the World Bank, in an act of 'green imperialism', encourages the growing of raw materials for these 'dirty' industries in the hills and forests of Andhra in the name of afforestation. The tribals who cultivate food crops or any other commercial crop of their choice in the so-called forest lands face severe repression from the forest department. The 'forest' lands have become so sacrosanct that the forest-dwellers are sought to be kept out of them. The only way they can continue to live in their own traditional areas and work on the lands which traditionally fostered their communities is by accepting the framework of World Bank-funded Joint Forest Councils wherein they have no right to raise the crops they need or prefer but are forced to raise bamboo, eucalyptus and other monoculture exotics. For accepting a meagre initial sum they will have to forgo their right over not just their lands but also on the World Bank-specified crops they have raised. They don't have an independent say on when they can cut the trees that they have raised, and to whom to sell them and at what price. Worse still, they don't even have the tenurial rights over these lands which are supposed to be jointly 'managed' by them. Here is a clear case of a state government unquestioningly sacrificing the interests of the tribals, for the sake of some meagre funds, at the altar of this institution of international finance.

In fact, the tribals are the most exploited and deprived section in Andhra. 1981 census figures say that the literacy among the STs in the State was 7.82% and among the ST women it was 3.5% The New Forest Policy cruelly pits the question of environment against the question of survival. It is not the tribals but the greedy vested interests from outside who are responsible for the depletion of forest cover. Armed by the Indian Forest Act, these tribals are systematically deprived of their rights over their traditional land. Land survey and settlement, which started under the British, was never carried out systematically in Andhra Pradesh as in most other states. Private ownership in land was virtually unknown among tribals and other non-tribal forest dwellers' communities earlier. While the landlords and the peasants in the plains had the benefits of various land settlement measures, not all the tribals got pattas for their land. It is the so-called civilised society, the class-divided society, which imposed its principle of private ownership in land among tribals only to deprive their control over their own land. Except in certain states in the Northeast, the tribal community's collective ownership right over community land is not even formally recognised by law. This principle of private property in land coupled with the Forest Act has wrecked havoc in the lives of tribals.

Of late, in Andhra Pradesh, there is an increasing clamour for revoking land transfer regulations protecting tribal lands. Cutting across party lines, both the Congress and TDP members were loudly demanding scrapping of these regulations right on the floor of the assembly. An earlier attempt by Chenna Reddy to prohibit eviction of non-tribals from the tribal areas was scuttled by High Court. On the other hand, the Supreme Court's insensitivity to issues of social justice was evident from the fact that the apex court has been sitting tight for the last 15 years over an appeal against AP High Court verdict which shot down a clause giving retrospective effect to Regulation 1 of 1970, a clause which could have benefited tens of thousands of tribals.

LAND ACQUISITION AND DEVELOPING LOCAL PARTICIPATION

Land acquisition and compensation are only two important issues as well. Since completely avoiding dislocation is sometimes difficult, the importance of carefully assessing the scale of disruption and preparation of a plan for resettlement can hardly be overemphasized. In India and elsewhere, nongovernmental organizations (NGOs) have been useful in mobilizing local resources and institution-building. The key to any successful resettlement programme lies in defining and developing mechanisms for people's participation in planning and implementation; this will further facilitate social cohesion and the development of a sense of community among the resettlers.

Indigenous Peoples and Water Rights

Till taught by pain,
Men really know not what good water's worth;
If you had been in Turkey or in Spain,
Or with a famish'd boat's-crew had your berth,
Or in the desert heard the camel's bell,
You'd wish yourself where Truth is—in a well.

Lord Byron (George Gordon Noel Byron)
Don Juan.

Water—the source and sustenance of all life—increasingly is becoming a source of catastrophic conflict and destruction. The successive hurricanes that battered the planet during the last 12 months, the melting of the polar ice caps, even the genocide in Darfur, are the fallout of unbridled human competition for resources, for good purposes and bad. As the human population and demand for development rise, the consequence of our failure to manage finite resources, such as water, in sustainable ways becomes ever more devastating. Indigenous peoples the world over remind us that the path humanity has chosen is not the only option. While their cultures and environmental circumstances differ vastly, indigenous peoples share an ethos for living in harmony with nature. They take pride in their respect for the earth and her resources. When forces outside their control compel them to adapt to changed conditions, they tend to innovate in environmentally sustainable ways. In both worldview and practice, they offer much to emulate.

In March 2006, the Fourth World Water Forum will take place in Mexico City. There nongovernmental activists, international lending agencies, and senior ministers of most of the world's governments will discuss ways that the target set in the United Nation's Millennium Development Goals—to reduce by half the more than 1 billion people who lack access to safe drinking water—can be met. This year participants will focus on solutions that have been implemented successfully at the community or local level, and that show promise for replication in other locales. This issue of the *Cultural Survival Quarterly*, which focuses on Indigenous Peoples and Water Rights, is timed to contribute to that discussion.

The Right to Water

To begin, it is worth reminding ourselves that water is a basic human right; indeed it is a prerequisite for realizing all other human rights, including the right to life itself. The Committee on Economic, Social and Cultural Rights, the body established by the United Nations to authoritatively interpret the meaning of the rights set forth in the International Covenant on Economic, Social and Cultural Rights (ICESCR), has declared that the "human right to water entitles everyone to sufficient, safe, acceptable physically and affordable water for personal and domestic uses." That right is available to all without discrimination of any kind.

In addition, the Committee calls upon States to give special attention to individuals and groups who have traditionally faced difficulty in exercising their right to water. In particular, States should ensure that:

Indigenous peoples' access to water resources on their ancestral lands is protected from encroachment and unlawful pollution. States should provide resources for indigenous peoples to design, deliver, and control their access to water. The burden falls on States to ensure that the right to water is respected. Not only must States refrain from interfering with the enjoyment of the right to water, they also must prevent third parties, such as large agricultural or corporate interests, from doing so. To ensure the right to water for future generations, States must adopt comprehensive and integrated water management strategies that may include:

- reducing depletion of water resources through unsustainable extraction, diversion, and damming;
- reducing and eliminating contamination of watersheds and water-related ecosystems by substances such as radiation, harmful chemicals and human excreta;
- monitoring water reserves;
- ensuring that proposed developments do not interfere with access to adequate water;
- assessing the impacts of actions that may impinge upon water availability and natural-ecosystems watersheds, such as climate changes, desertification and increased soil salinity, deforestation and loss of biodiversity; and
- establishing competent institutions and appropriate institutional arrangements to carry out the strategies and programmes.

These are the precise sorts of measures that indigenous peoples have called for in the Indigenous Peoples Kyoto Water Declaration, which they delivered at the Third World Water Forum in Kyoto, Japan, in 2003.

Violations of the Right to Water

Throughout the Andes, in India, and in the Philippines, indigenous peoples are facing increasing competition for their scarce water reserves from agricultural plantations, as well as hydroelectric, mining, and drinking water companies. In Chile, the government's privatization of water, combined with its failure to provide its indigenous peoples with timely and adequate information about how to register their water rights, sold their right to water out from under them. Similar abuses are illustrated from other parts of the world. On Black Mesa, the ancestral homeland of Hopi and Diné (Navajo) peoples in the American southwest, mining by the Peabody Coal Company has nearly depleted the aquifer on which Native Americans rely for their drinking water. Nearby, peoples from several indigenous nations have gone to court in an effort to stop a ski area from spraying reclaimed wastewater to make artificial snow on the sacred San Francisco Peaks. In the Pacific northwest, mercury from abandoned gold-rush era mines and PCBs and other contaminates dumped into rivers by the US military have polluted the water supply and made fish—a staple of the traditional diet—unsafe for human consumption.

Though not specifically addressed in this issue, CSQ has covered the impact of hydroelectric dams and other large-scale development projects in past issues. More recently a study, carried out by the United Nations Special Rapporteur on the Situation of Human Rights and Fundamental Freedoms of Indigenous Peoples, found that where such developments occur in areas occupied by indigenous peoples, it is likely that their communities will experience profound and unforeseen social and economic changes that devastate their ways of life. Cultural Survival has previously reported on the situation of the Embera-Katio in Colombia, a portion of whose ancestral territories was flooded without their consultation in 1992 to build a hydroelectric dam. As a result, they were forced to move into an area that is heavily ravaged by the country's ongoing armed conflict. Many of their homes and property have been destroyed, several of their leaders have been assassinated or forcibly disappeared, and many have been displaced for a second time as a consequence.

It is no exaggeration to say that violation of the right to water can lead to violent conflict. In Bolivia, for example, the government, at the behest of the World Bank, turned over management of the Cochabamba city water and sewage system to a single-bidder concession of international water corporations in 1999-2000. Under the arrangement, which was to last for 40 years, water prices increased immediately from negligible rates to approximately 20 percent of monthly family incomes.'

Citizens' protests were eventually met with an armed military response that left at least six residents dead. The protests continued until the consortium was forced to flee the country. Of more immediate concern is the situation in Darfur, Sudan, where decades of drought and desertification have intensified competition for scarce resources, particularly water. State interference with traditional dispute resolution mechanisms, and the superimposition of highly corrupt state administrative agencies, compounded underlying intergroup tensions. Over time, these flared into anti-state sentiment that turned violent as a result of the rebels' easy access to modern weapons. With its attention taken up by peace negotiations in the south of Sudan, the government turned to its supporters among Arab nomadic herders to act in its place.

In short order this government-backed contingent, who lacked land and were looking to settle, became a marauding force, the Janjaweed. The Janjaweed has targeted the civilian population for infamous abuses that include mass murder, the burning of villages and displacement of entire communities, and the rape of women and girls. Nearly a year after massive international attention, the situation refuses to abate.

LAND ACQUISITION AND COMPENSATION

The consequences of forced resettlement are enormous: it destroys the existing modes of production and ways of life, affects kinship and community organization and networks, causes environmental problems and impoverishment, and threatens cultural identity of tribal and ethnic minorities, often the worst sufferers of hydroelectric dam projects. The forced resettlement of populations in association with the construction of hydroelectric dams, large-scale irrigation projects, highways, mines, and urban renewal and metropolitan development has become commonplace today. The extent and the implications of such forced relocation are diverse and variable, depending on the nature of the project and density of population being affected. The size of a displaced population being affected. The size of a displaced population may vary from only a few thousand to tens of thousands of people.

The Kaptai hydroelectric project, for example, displaced close to 90,000 slash-and-burn agriculturalists in the Chittagong Hill Tracts in Bangladesh. The Aswan Dam project in Egypt resettled more than 10,000 people; the Narmada Dam project now under construction in Gujarat, India, will displace some 70,000 people; a recently completed dam project in Togo displaced 10,000 people; about 60,000 urban and rural people were displaced by the Sobradinho Dam in Brazil; 383,000 people were forced to resettle due to the inundation caused by the Danjiangkou Dam project in China; and the Shanghai Metropolitan Development Project in China and the Dhaka Water Supply and Sanitation Project in Bangladesh now underway will cause relocation of 50,000 and 20,000 people, respectively. In India alone, according to some estimates, between 1.5 and 2 million people have been displaced by various development projects since the 1950s. Given the nature of rapid urbanization, industrialization, and agricultural modernization in many developing countries, there can be little doubt that the need for involuntary resettlement will increase in the near future.

In addition, forced resettlement tends to be associated with increased sociocultural and psychological stresses and higher morbidity and mortality rates. Population displacement, therefore, disrupts economic and sociocultural structures. People who are displaced undergo tremendous stress as they lose productive resources-land or otherwise-in the adjustment process. Resettling the displaced poor, remote, and economically disadvantaged is not always an easy task.

To date, social science research has largely focused on impacts of dislocation and adjustment strategies. Two previous issues of Cultural Survival Quarterly also centred on problems of dislocation and resettlement. This article primarily focuses on one important aspect of the resettlement process: land acquisition and compensation. What should be the contents and scope of a resettlement plan? Should the displacees be involved in the decision-making process? How do you protect the interests of vulnerable groups such as tribal people, landless and seminlandless peoples, and women groups? How do you stop the influx of nonresidents from taking advantage of compensation arrangements? Should there be a national resettlement and compensation policy?

Most countries have land acquisition laws that require prompt and adequate monetary compensation for persons who lose their land and property. However, cash compensation has many negative consequences, particularly for tribal and other marginal populations. Tribal economies are in large part non-monetized, based on reciprocal exchange of goods and services; therefore, people are not well accustomed to managing cash. There is a popular saying among the Havasupai Apache Indians in the United States, a people displaced repeatedly by development projects: "Land is like diamonds but money is like ice".

Some tribal people have very little transaction experience with the outside world. For example, the cash compensation given to those affected by the Kulekhani hydroelectric project in Nepal, which displaced 3,500 tribal people, produced a greater impact on peoples' behaviour, in many instances, than did loss of property:

The sudden cash in their hands gave many the false impression of wealthiness. They changed their life style. The Tamangs of Markhu, who despite their fight against the land acquisition, had lost the best part of their land, became the living example of this side effect [of cash compensation]. Gambling and drinking increased to an unprecedented level.

Eight male adults developed severe tuberculosis and all but one died in a miserable condition in the following five years. One young Tamang who had received a large sum in compensation went as far as Casino Nepal in Kathmandu to gamble until he lost all his money and ended up working as a dishwasher in the city... The story of damage ones on and on... The villagers...think that displacement has cost them a lot both materially, and spiritually. They claim that they were not compensated adequately for their loss. There were serious flaws in compensation planning in the Kulekhani... [that it has become] increasingly important that we review and improve our compensation policy.

Another project in Indonesian revealed that "displaced families provided only cash compensation suffered about a 50 percent reduction in income compared to pre-project conditions, and their productive resource base was reduced by 47 percent". Studies carried out at the Srisailam and Lower Manair dam projects in India also confirm the inadequacy of cash as a mode of compensation. Exclusive cash compensation is least useful to the resettlers in the long run; instead, for land-based resettlement, a "land-for-land" approach may be more beneficial since land is the key to reestablishment and contributes to cultural security. Guggenheim (1990) considers that the land-for-land approach is essential for resettlement despite criticisms that it denies the settlers the right to choose for themselves and that cash compensation could provide opportunities for opening family businesses.

One important mechanism for implementing the land-for-land strategy is to identify several possible relocation sites to provide alternative choice to the displacees. The productive potential, quality of soil, availability of irrigation water, and locational advantages of the new relocation sites should be ideally better or at least equivalent to the lost site in order to make it attractive to the settlers. Furthermore, in selecting relocation sites, attention should be paid to the possibility of off-farm income to supplement family income. Partridge (1989a) notes that there are three significant issues related to monetary compensation: (1) evaluating the worth of property to determine the amount of payment; (2) the timing of the payment; and (3) determining noncash compensation where cash alone is not appropriate. In many countries, market value of the land being acquired is used as the determining factor in calculating compensation. A displaced person may find it difficult to acquire comparable land with the compensation money because of limited land market or higher value of land in the relocated area, where prices can double or even triple almost overnight. In addition, the costs for relocating, transporting, salvaging building materials, and so on can put financial strain on the resettlers. In such circumstances, cash compensation should be supplemented by providing "replacement assets" in order for the displacees to be resettled.

Compensation payments are sometimes severely delayed-as much as 10 to 15 years in India, Bangladesh, and Nepal-which obviously devalues the compensation award. Compensation money must be made available before the actual move so that displaced households can use the money to overcome or minimize the hurdles of dislocation. Furthermore, compensation for land acquisition should not be limited to monetary payments to individuals; there should be appropriate compensation to the community of people to reestablish their new communities:

The Narmada Water Dispute Tribunal (NWDT), formed by the Supreme Court to adjudicate sharing of water of the Narmada river among four states in India, established in a land mark decision that property holders displaced due to land acquisition were entitled to cash compensation. The NWDT specified that irrigable agricultural lands, houseplots, schools, community buildings, drinking water wells, fuelwood lots, livestock wells, health clinics, threshing grounds, burial or cremation grounds, and so forth were to be provided by the government to replace those sacrificed to planned development projects. Only the cost of agricultural land is to be recouped from the cash compensation of the affected people; other components are project costs.

Finally, cash compensation disproportionately benefits some interest groups (*i.e.*, big landlords) and not so much poor and small-scale farmers, the landless, and women. Landlords profit more from relocation because much of their land is either sharecropped or remains uncultivated, making the cash compensation more attractive to rich farmers, who then reinvest it in the nonagricultural sector. Landless labourers are often the hardest hit group in the relocation processes due to their lack of ownership and entitlement to land. In many countries such as Bangladesh, India, and Pakistan, landless people constitute 50 percent of the total rural population. In such circumstances, there is often customary recognition of "use-right," but cash compensation to landless displacees is always inadequate. As a group, women are also affected by forced relocation; very rarely have their concerns been considered in resettlement projects. In many cultures, women are involved with land-based activities and in herding animals; they are part of the productive work force and contribute substantially to the sustenance of the family. However, compensation monies always go to men, often leading to mistrust and division in the family. Areeparampil (1989) has documented how some men, often hand in hand with corrupt officials, have fraudulently deprived indigenous people in Chotanagpur, India-particularly the weaker ones such as widows and elderly women-of their rightful compensation.

LAND TENURE, ACQUISITION, AND RESETTLEMENT PLANS

Involuntary resettlement plans must not only address the relocation and rehabilitation needs of property owners, but also the landless, wage labourers, artisans, craftspeople, and women. In other words, resettlement designs must keep the community in mind, not the individual. This means carefully considering existing land tenure systems and inheritance patterns, and local customary practices that govern use and entitlement to community property. Studies of land tenure systems and customary practices will be helpful in devising compensation rules and resettlement titles. Furthermore, developing and designing a resettlement plan means understanding the legal framework. Many World Bank-financed projects in Asia and elsewhere dealt with the problems associated with relocation and resettlement on an ad hoc basis due to lack of national resettlement policies. Since the legal basis for land acquisition may vary from state to state and region to region (as in India), a national land acquisition and resettlement policy should be flexible enough as an instrument for guidelines, leaving the details to local authority for local decisions on the merits of individual projects. Planning resettlement must begin early on in a development project.

It should be based on information as accurate as possible about the scale of displacement, impacts and consequences on the life and livelihood of the people, the extent of loss of assets, and destruction of infrastructure and services. A good strategy is to undertake a complete socioeconomic survey of potentially affected families, with names of all members of the household, their age, gender, education levels, and occupational backgrounds. The survey should also include questions aimed at gauging people's attitudes, preferences, and choices on future resettlement. Such a survey is of fundamental importance to plan ahead and to prevent inflows of outsiders for future compensation award.

Due to the lack of a proper survey, for example, the number of families requiring relocation almost tripled from initial estimates in the Upper Krishna Irrigation Project in India, causing considerable delay in resettlement operations. Participation of and assistance by the local community in the survey and planning process may ensure better results. Instead of a "top-down" resettlement planning, if the people are consulted right from the beginning and feel involved in the local decision-making process, they will be more likely to see the resettlement plan as their project. Partridge and Salam (1990) conclude that people's participation in resettlement planning and implementation can be very helpful in designing realistic and sustainable economic production systems. Finally, the socioeconomic survey data can also be used to develop new training programmes for alternative employment if the available land is not sufficient to accommodate all the project-affected people.

POLITICS OF LAND ALIENATION AND PROBLEM

Tripura exhibits classic case of land alienation where unchecked migrants across the international border completely overpowers land, economy and polity of the state. Marginalization of the Boroks shows the relevance of state intervention to protect the right and dignity of the innocent indigenous communities against the populous and advanced communities. Over time, land alienation has turned out to be the root cause of ethnic conflict and violence. While owning land can be a source of productivity and prosperity; losing the same could be source of ethnic conflict and human rights violation. The state of Tripura in the northeastern region of India represents a glaring example where the above problem is visibly and distinctively witnessed. Pursuing this problem the paper proposes to address to the following questions: Who are the indigenous people in Tripura? Have these historical communities in the state been dispossessed of their land? How did they eventually lose their land? How the increasing immigration is resulting in the alienation of indigenous peoples' land? And what are the consequences?

Tripura has been facing distinct socio-economic, political and ethnic problems, which are essentially driven by long spell of immigration from across the international border. The state covers an area of 10,491.69 sq. km. and shares border with Bangladesh in the north, west and south, and Assam and Mizoram in the east. The economy of the state is predominantly dependent on agriculture. Tripura has a unique history in the country where the original inhabitants – the indigenous peoples called the Boroks of Mongoloid origin – have been outnumbered by the influx of immigrants from across the international border of erstwhile East Pakistan, and now Bangladesh. To a great extent, the socio-economic and political problems in the state are related to one crucial process that has been happening unabated – the land alienation of the indigenous people.

Tripura has a history of being ruled by as many as 184 kings belonging to the Boroks. It is the homeland of about 19 sub-clans of indigenous peoples who share common features and identity as Borok peoples. The Boroks comprises of many sub-clans namely Debbarma (Tipra of Tipperah), Reang, Jamatia, Koloi, Noatia, Rupini, Halam, Hrangkhawl, Ranglong, Kaipeng, Tripura, Morasing, Molsom, Darlong, Mog, Bongcher, Chorai, Uchoi, and the likes. Except one or two of these sub-clans, all of them speak a common language called the Kokborok. In fact, the name "Tripura" is a corrupted form and is derived from Kokborok word, "Twipra." The name "Twipra" is a combination of two Kokborok words– *Twi* (water) and *Bupra* (tributaries). It is widely consented that the Boroks migrated from parts of China, and in the process, reached the present territory, which was not occupied by any ethnic group or people, and started settling nearby the areas where plenty of water was available. So the word "Twipra" indicates the Boroks as the indigenous peoples, owner, first occupants and son of the soil of the state. The Boroks were known to have practiced shifting cultivation as the main traditional source of economy and means of their livelihood. Misra in his book, *Socio-Economic Adjustment of Tribals,* referred them as *jhumias,* which applies to those indigenous peoples who practice shifting cultivation like other tribes of Northeast India and Chittagong hill tracts of Bangladesh.

To speak about the political history of Tripura, it had been a princely state until it was completely merged to the Indian Union on 15th October 1949 after which it went through a series of political status – it first became a Territorial Council in 1956, then a Union Territory in 1963, and attained statehood on 21st January 1972. Today, the largest threat to indigenous population comes from the unabated influx of peoples mainly from across the Bangladesh border. The beginning of this influx can be linked to the time of India's partition, when the Indian state showed indifference to the idea of protecting the territorial integrity and identity of the indigenous Tripuris. This historical blunder reduced the indigenous peoples into a minority in the land and territory of their own.

In Tripura, the growth of population has been so phenomenal that over a decade following the partition of India, the territory registered more than twenty times increase in its population. After the partition, in the decade from 1949 to 1959 alone the total number of influx of refugees from across the international border figured 3,73,500, and in the year 1964–65 itself 1,00,304 refugees migrated to Tripura. The total number of recorded or registered refugees within the period from 1947 to 1971 was 6,09,998. As a result, the immigrants began to outnumber the Boroks since 1960s thereby reducing the later into a minority. The Boroks, whose population were once more than 95 per cent, now constitute only about 30 per cent of the total population of the state. Today, the immigrants had not only outnumbered the Boroks but also marginalized them and dominate over the economic, political, administrative and social spheres of the state. This process naturally make the Boroks insecure, creating a sense of discontentment and denial of their basic rights to self-determination. Traditionally, the Boroks practiced jhuming or shifting cultivation because not only the technique of settled cultivation was unfamiliar and contrary to their tradition, but also they found the hill slopes unsuitable for plough cultivation. Gradually, influx of population and scarcity of space compelled the Boroks to shift from shifting cultivation to settled cultivation. At the same time, the latter was prominently emphasized as a desirable practice in India's post-independent period. The scarcity of land was largely created by the unchecked large-scale immigration. Such a form of immigration had significantly affected the Borok's economy. This led to Borok people's alienation over their own land and also to the non-tribal immigrants in the state. Some people identify alienation of land as the root cause of socio-economic problems, political unrest and ethnic conflicts in the state. The unceasing influx of "foreign nationals" had deprived the indigenous peoples' rights over the lands and employment opportunities. The settlement of large number of immigrants in both the plains and hills of the state has led to (i) displacement of the Boroks, (ii) non-availability of land for rehabilitation to jhumias, and (iii) non-availability of land for distribution among landless local peoples.

Going by the tradition of shifting cultivation, land was owned collectively by the community, and the rights of ownership enjoyed by an individual local person over a particular area of shifting cultivable land was neither permanent nor alienable nor transferable to any person nor inheritable from any person. Prior to the partition of India, as the density of population was very low in Tripura, the foodstuff required by the Boroks was minimal and could be met easily from shifting cultivation. As such the issue of transfer of lands did not arise, and therefore, the problem over land alienation was non-existent.

Each Borok family had huge area of wetland under their possession. Although they were not using those lands all the time, the same were used from time to time for jhuming. To induce his subject (Boroks) to settled cultivation and to increase state's revenue, King Bir Bikram Kishore Manikya (1923–47) reserved for the Boroks a large area of fertile cultivable lands by order in 1931 and 1943 covering the areas of 110 sq. miles and 1,950 sq. miles respectively. Thus, technically speaking, all the fertile cultivable lands were made reserved for the King's subjects and many families were settled in reserved areas and taken to settled cultivation by 1940s. Further, as a measure to protect his subjects the last King of Tripura declared that no land in the reserved areas could be transferred to others without the permission of the Government. But despite this declaration and safeguard, the reserved areas lost its balance, mainly after the partition of India (1947) and Tripura's merger with the Indian Union. Thereafter, the rate of inflow of immigrant population increased sharply forcing the state authorities to utilize the reserved areas for building refugee colonies.

Describing the impact of large-scale of immigration in the state, Malabika Das Gupta opines that "[w]hile there has been a general lack of economic development in the state, by and large, the tribal people of Tripura have been the worse sufferers as a result of this phenomenon." The massive influx of refugees into the state, which had made the indigenous people into a minority in the land of their own, has added fuel to the fire and accentuated the alienation of Borok peoples' land despite legislative measures to prevent the same. Besides, the immigrants claim to have brought with them the superior agricultural technology of plough cultivation. Though a handful of Boroks had tried to adopt plough cultivation, the majority of them lost their plain lands to the non-tribal immigrants and was compelled to retreat into interior hills and forests of the state. Many of the Bengalis immigrants had taken to trading and money lending and steadily the Boroks came under the exploitative influence of the immigrant moneylenders and traders.

The failure of the jhum products and revenue to meet the growing needs of the Boroks in a way led to their dependence on non-tribal moneylenders and businessmen for loans were used to buy necessary items they could not produce. Loans with interests to be repaid within fixed periods of time were given to the Boroks only against the mortgage of their plains cultivable lands. Meanwhile the moneylenders could use mortgaged lands for cultivation. Mortgaging of cultivable land led to dwindling of production for the Boroks, and they fail to repay the debts within the stipulated time. The immigrant moneylenders and traders caught the Boroks in a debt trap of loan capital. Failure to pay interest within the stipulated period enabled permanent possession of the mortgaged lands by the moneylenders. Instead of settled cultivation giving prosperity to the Borok farmers, they found themselves victim of money lending system. The entire process led to alienation of the Boroks from their own lands, which further led to their general pauperization.

The condition of the Boroks worsened with the handling of "Reserved Areas" in the state. The Government of Tripura prohibited large-scale jhuming since 1952 and restricted the customary rights of the Boroks and their free movement in the forest. Even very little wetland is left for the Boroks to practice settled cultivation. The Government policy deprived the Boroks of their sources of livelihood, as they were not provided with a viable alternative to improve their condition.

As a measure to check alienation of the Boroks, a special provision was included in the Land Revenue and Land Reforms Act, 1960. Technically the Act imposed restrictions on transfer of land from Boroks to non-tribals. Section 187 of the Act provided that "no transfer of land by a person who is a member of a Scheduled Tribe to a person who is not a member of such tribe, shall be valid unless the transfer is made with the prior permission of the Collector in writing."

However, this Act was inadequate to prevent the transfer of Borok lands to the non-tribals. There were rooms for administrative manipulation embedded in the Act. By avoiding all the legal impediments, the immigrants began to purchase lands from the Boroks through higher rate and Benami deeds. For instance, if the value of a plot of land among the Boroks is Rupees five thousand, a non-tribal buyer offers a sum of Rupees ten thousand. Consequently, the Borok sells the land to the non-tribal under a fictitious tribal name in paper. It is evident from such Benami deeds that the motive of the non-tribal immigrants is to occupy the Borok land by all means.

Although the Land Reforms Act, 1960, prohibited transfer of land to the non-tribals without prior permission of the Government, but in practice, due to its ineffective implementation, indiscriminate transfer of land from the Boroks to the non-tribals took place in contravention of the legal provisions. In a Block level survey of indebtedness and alienation of land conducted by the Directorate of Tribal Research, Tripura Government, Agartala in 1974, it was found that indebtedness was directly and positively correlated with alienation of Boroks' land. In most cases it was revealed that heavy load of indebtedness particularly after drought years of 1965 and 1972 forced the Boroks to sacrifice most part of their lands.

The Land Reforms Act 1960 could not prevent the transfer of Borok lands to the non-tribals. The Act was amended in 1974, which provided for the restoration of all Borok lands transferred on or after January 1969, in contravention of the Act, to the transferor or his successor-in-interest. By keeping the lands transferred before 1st January 1969 outside the scope of the Land Reforms Act, legal recognition was given to the transfer of lands made before that date.

The Tripura Land Reforms (Third Amendment) Act 1975 contains the provision that "if any non-tribal people intends to sell land within 42 Tashils and 133 villages specified in the Act, then the members of the Scheduled Tribe shall have the preferential right to purchase such land." However, in practice, there is hardly any record of non-tribal community selling lands to the Boroks. At the same time, the Boroks being economically much backward than their counterpart are not in a position to buy back land from the non-tribal immigrants.

Notwithstanding the several Land Reforms Acts passed by the Government of Tripura since 1960, the plight of the Borokswith regard to the illegally transferred lands could not be minimized at all. Though the law may provide some provision of protection, the fact is that the rural Boroks are generally poor, illiterate and have no capacity to access and knowledge about judicial redressal alternatives. It is appropriate to quote one of the observations and reports submitted by the Dinesh Singh Committee constituted by the Government of India to suggest measures for restoration of peace and normalcy in Tripura after the violence of June 1980. It says: Two Features if this Tripura Land Revenue and Land Reforms Act caused dissatisfaction... Firstly, with regard to the choice of the end of 1968 as cut off point, the Acts virtually legalized the transfer of Boroks' lands that had taken place before January

1969. Secondly, the progress of restoration has been lamentably tardy. It received some impetus in 1978 part and has again languished thereafter. Out of nearly 100000 petitions received only 4300 have been disposed off in the petitioners' favour and just 7700 acres of lands have been actually restored till March 31, 1980. Though, the above observation seems to reveal the restoration of Borok land, but in reality it was not completely materialized. The Boroks continue to remain alienated from their own land.

The question of land alienation of the Boroks and its restoration has all along been pertinent in the context of the political scenario of Tripura. This issue received importance in the ideologies of various socio-political organizations of the indigenous peoples of the state, like the Tripura Upajati Juba Samity (TUJS) and Tripura National Volunteers (TNV). One of the main demands of these two organizations was the complete restoration of illegally alienated lands. However, the TUJS failed to achieve this objective when the party was in power during the coalition government in the state for a complete term of five years in 1988. On the contrary, the TNV was persistent on the question of land alienation and its restoration. As a result, the issue was incorporated in the TNV accord of tripartite agreement signed on 12th August 1988. As per this Memorandum of Settlement (MoS), Section 3:6 contains the following agreed resolutions:

1. Review of rejected applications for restoration of Boroks' land under the Tripura Land Revenue and Land Reforms Act, 1960.
2. Effective implementation of the law for restoration; and
3. Stringent measures to prevent any fresh alienation.

Though the MoS was signed by a tripartite agreement between the TNV, the Government of Tripura and the Union Government, restoration of lands to the Boroks has not yet been implemented. Subsequently, TNV has threatened the government with renewed agitation in case the Accord is not respected. The surrendered members of the TNV even declared that they would again go back to armed struggle if the provisions of the TNV accord were not fully implemented. Renewed demand for implementation of several Land Revenue and Land Reforms Acts has received wide range of support from TNV, TUJS, regional political parties, Tripura Students' Federation (TSF) and all sections of the Borok society.

The Government of Tripura came under pressure from various sections of the Borok society in 1994-95. CPI (M) led Front Government in Tripura under the leadership of Borok Chief Minister, Dasharath Debbarma, initiated certain measures to implement the restoration of alienated lands to the Boroks. The Left Front Government passed the Tripura Land Revenue and Land Reforms (Sixth Amendment) Bill 1994 and declared 1994–95 as the year of land reforms. This bill aims at reinforcement of all provisions of Land Revenue and Land Reforms Acts of 1960, 1974 (as amended), 1975 (as amended) and 1982 (as amended) particularly relating to the restoration of lands to the Boroks.

It also incorporated the provision of prevention of re-transfer of restored lands to the non-tribal people. The Act provides the provision to safeguard the interest of the indigenous peoples who otherwise had become victims due to the massive influx of refugees. As per the provisions of the Act, the Government initiated to implement the law to restore all the alienated lands to the Boroks from the non-tribal people. In many areas the non-tribal people who possessed illegally transferred lands were issued notice and order to return the lands back to the original Borok landowners.

However, the bill faced great opposition from the non-tribal communities even before its implementation in the state. They were determined not to return the lands back to the Boroks irrespective the legal directives. A regional political party, Amara Bengali, strongly reacted against the bill. This phenomenon created ethnic tension between the Boroks and the non-tribals and led to several clashes in different parts of the state. It is reported that the non-tribals were organizing themselves to protest against the implementation of the Land Revenue and Land reforms Acts 1994. However, such a strong opposition is possible only in a state like Tripura because of the imbalance in demographic composition.

Following these developments, restoration of land related tension was mounting high at Teliamura area in West Tripura following the massacre of six members of a non-tribal family on 29th April 1994. As the people of Thangsai in Telianmura area of West Tripura District became panicky and insecure after the incident, the state government had to take measures to establish confidence among the people, and security forces were kept in close vigil. The massacre occurred within a week of the state government's declaration of 1994–95 as the year of land reforms, which indicates opposition to the law relating to restoration of Boroks' land. The ethnic tension had forced the Tripura government to issue an order of special alert to all the police stations, sub-divisional and district heads of police to maintain a maximum vigil on the prevailing situation and take measures to prevent any communal violence in the state. As the tension was mounting high, the then Chief Minister, Dasharath Debbarma, through telecast message on 9th May 1994 appealed the people of different communities to maintain peace and communal harmony in the state. According to the Director General of Police the situation at that point of time in the state was extremely serious, and showed reminiscence of the 1980 riot.

The problems arising out of land alienation and its restoration is due to the negligence of the Government of Tripura to implement several Land Revenue and Land Reforms Acts. If the Land Reform Acts passed since 1960 are not implemented fully and sincerely, the future of the Borok peoples and their economy cannot be safeguarded. A five-member TNV political party delegation lead by its President, B.K. Hrangkhawl, met the then Chief Minister, Dasarath Debbarma on 10th May 1995 and submitted a memorandum containing five-point charter of demands. The memorandum included the demand of deportation of "foreign nationals" and physical restoration of alienated lands, both of which were included in the provisions of the tripartite TNV accord signed on 12th August 1988. The TNV meanwhile had threatened to paralyse the entire state if its demands are not conceded and fulfilled on time. It is quite clear that alienation of the Borok's land and its restoration assume a significant political dimension. There is a need for a government with a strong political will to resolve such an on going problem in the state. If the issue of restoration of land is not tackled properly on time, it appears that ethnic conflict and communal disharmony may aggravate in the days to come.

Government of Tripura faces a stiff challenge to stop land alienation of the Boroks as this directs causal linkages with the ongoing socio-economic problems and ethnic conflicts. But the question remains if the government is willing to resolve the problem or is only wary of majoritarian electoral interests. Such a concern has hardly appeared in the imagination of the dominant political class in the state. In the face of impending crisis of land alienation it has to be said that the sooner the process initiated the better it will be for the people and the state. Any approach towards this should recognize that simply making legislations after legislations is not going to take much way forward if not guided by a strong will to uphold the rights and historicity of the indigenous people. One way of overcoming the crisis could be through educating the opposing masses with right sense of understanding the issue. Secondly, creation of viable alternative is extremely important for those who would give back the lands, for they too need security and alternate source of livelihood. Thirdly, there is a need to create economic opportunity for all sections of the Borok people without any political favour. Such initiatives will certainly instill a proper understanding and good will among the people of the state, which will perhaps ensure a better Tripura ahead.

LAND ALIENATION AND PAUPERIZATION

The colonial state imposed the land laws based on this worldview on the traditional communities. This imposition affected the dominant castes as well as the tribal communities. But many of the former had access to education and other modern inputs. So they had some preparation to deal with the changes. Most traditional tribes, on the contrary, lived on mineral and forest rich land that the colonialist required as raw materials. That turned the imposition of the formal system on the informal societies into an unequal encounter. Land alienation from the traditional to "modern" communities was a consequences since the latter were unable to deal with the changes imposed on them.

This unequal encounter continues to be the basis of a disjunction and of conflicts between the two systems because the colonial laws continue to be in force in the country. One of its consequences is environmental degradation. The legal system that recognises only individual ownership is a major cause of land loss and environmental degradation. Since the CPRS are not recognised as their sustenance, the communities depending on them cannot prevent outsiders from encroaching on that land. For example, in Tripura in North Eastern India, the tribal proportion has declined from 58 percent in 1951 to 31 percent in 2001 because immigrants have encroached on 60 percent of their community owned land with the help of individual-based laws. Equally important is loss of forests which catered to many needs of the tribal and other rural poor communities. The state handed many of them over to industry as raw material. They were treated as sources of profit and destroyed with no concern for their dependants or for conservation. That impoverished people (Gadgil 1989.

The third source of land loss is acquisition for development projects. The law that empowers the state to acquire land recognises only individual ownership. More than 25 million hectares have been used in India for such projects 1947-2000, around 14 million of them forest and other CPRs. Their inhabitants, most of them tribal and other rural poor like fish and quarry workers are considered encroachers and are not compensated and often not even counted among the displaced (Fernandes 2008: 92).

Often records of the CPRs are not kept since they are considered state property and their inhabitants are encroachers. For example, according to official accounts, in Assam the state used 159,017.37 hectares of land for development projects and displaced 343,262 persons from them 1947–2000. The reality is 1.9 million persons displaced from 567,281.29 hectares (Fernandes and Bharali 2006: 107). More than 1.5 millions displaced persons and 410,261 hectares were not counted because according to the law these CPRs are state property and their inhabitants are encroachers with no right to live there.

The Vicious Circle

One can mention many other modes of land alienation. The above examples are given only to show the processes that lead to alienation of the people's livelihood. Because of the unequal nature of the encounter, also the reaction of these systems to the problems that the process causes differs. That too is based on their worldview on land and the natural resources all of whose dependants feel the negative impact of the transition from the traditional to the modern economy and new values. But the rural poor, particularly the tribes and other forest dwellers feel its impact more than the remaining groups do because it is an attack on their tradition of judicious use of resources and on the systems they had developed to manage land, forests and CPRs as their renewable sustenance. Loss of their sustenance begins the vicious circle of impoverishment that forces the dependants of these resources to overexploit them and cause further environmental degradation and more poverty. As the former Brazilian President Fernando Henrique Cardoso (1998) said, the first danger to the environment from people's impoverishment is loss of biodiversity and linked to it, loss of the values through which the communities depending on it had managed the resource as renewable. Studies show that loss of this value system or ideology is basic to the vicious circle that leads to further environmental degradation.

But reaction to this process differs according to the class one belongs to and one's ideology or culture. To the urban middle class land alienation and environmental degradation are loss of their recreational spaces while to the rural, particularly tribal, communities it results in loss of their livelihood and consequent impoverishment from which follows further land alienation and destruction of more natural resources. Conflicts are a natural consequence of this contradiction.

The first step of this process is impoverishment of the economic status they are reduced to by the alienation of their sustenance. It begins with landlessness. Then comes joblessness. For example, studies of families displaced by development projects show that in Andhra Pradesh in South India, the proportion of the landless rose from 10.9 per cent before the project to 36.5 per cent after it and in Assam in the Northeast from 15.56 to 24.38 per cent. Even among those who retain land, the average area owned declined, for example in Assam from 1.2 to 0.6 hectares per family. In every state most small and marginal farmers became landless, and medium farmers joined the ranks of small and marginal farmers. They also witnessed a decline in the support mechanisms such as the number of irrigation ponds and wells, poultry, cattle, and draught animals that used to supplement their agricultural income declined.

Joblessness is the next step. The land and other resources that are alienated from them used to provide them work. They lose this resource with no alternative to take its place. Joblessness resulting from it takes two forms. The first is lower access to work and the second is downward occupational mobility. In Andhra Pradesh, for example, 83.72 per cent of the land losers used to work on their land or elsewhere before its loss. After land loss access to work declined to 41.61 per cent. In West Bengal it declined from 91.02 to 53.18 percent and in Assam from 77.27 to 56.41 percent. The second is downward occupational mobility. In most states, more than 50 percent of the land losers who were cultivators before it became landless agricultural labourers or daily wage earners after land loss.

Also displacement can continue as a result of environmental degradation. For example, a new industry often forces people to move out of its neighbourood because after its construction environmental or other consequences such as fly ash and dust generated by the thermal, aluminium, nuclear, cement and other plants destroy the land around it and render it unusable. Its dependants cannot sustain themselves on it and are forced to move out. Also the noise and dust pollution and constant blasts in the coalmines often force people to leave their homes.

Absorbing a New Culture

The changes do not remain external but enter the community itself through the internalisation of the dominant culture. The major change is the culture the community in general and its elite in particular internalise, of viewing their sustenance as commodity alone. It is seen firstly in the demand the leaders make that individual ownership become the norm in their communities. For example, in the Garo tribe of East Garo in Meghalaya in Northeast India the leaders accepted the culture of individual ownership in the 1980s. A study two decades later shows that 30 percent of the tribal families in this district had become landless since their elite had monopolised much of their land.

These changes also have gender implications. As stated above, even the matrilineal societies are patriarchal. Their leaders absorbs the culture of greater patriarchy and express it in their land relations. That can be seen among the Garo who are a matrilineal tribe but individual ownership is through men. Among the Khasi of Megalaya who too are a matrilineal tribe, the male leaders who control the village council exploit their power to their own advantage and turn community-owned land into their private property (Mukhim forthcoming). Such change of gender attitudes is seen in other tribes too in the manner in which men interpret their customary law and property relations in their own favour.

Communities thus deprived of their resources absorb the same culture in another form. The first is the vicious circle of viewing their resources as a commodity alone. Once they are deprived of their resource and are impoverished, for sheer survival they overexploit the same resource for an income. For example, studies in all the tribal areas show that once they lose their land, the deprived families fall back on their forests that they had preserved for centuries and cut trees for sale as firewood or timber, and cause more deforestation.

The second to view their own bodies as a commodity. For example, 49 percent of the families displaced by development projects in West Bengal and 56 percent in Assam pulled their children out of school in order to turn them into child labourers. Women began to view their bodies only as a source of income. Because of it prostitution grew enormously among the families that had lost their land. All these instances point to a major change in subaltern culture. These communities lose hope in their future and think only of the present. As a result, children who are an asset for the future become commodities only for the present and are used as a source of income for survival. The same is the view of women's bodies. In other words, women and children become commodities more than men do.

This chapter which is an overview of the changes in land relations, has shown the new culture that grown as their result. It shows that imposition of another culture on a traditional group can result in a culture that is destructive of a community in general and of women in particular. The solution is not either going back to their tradition by opposing modernisation or absolutising the modern system. One cannot prevent all individual ownership either. One has to find an alternative in beginning with the tradition community values and combining them with the traditional community. Tradition has to be modernised and not replaced completely.

LAND USE, LAND-USE CHANGE AND FORESTRY

Land use, land-use change and forestry is defined by the UN Climate Change Secretariat as "A greenhouse gas inventory sector that covers emissions and removals of greenhouse gases resulting from direct human-induced land use, land-use change and forestry activities."

LULUCF has impacts on the global carbon cycle and as such these activities can add or remove carbon dioxide from the atmosphere, contributing to climate change. LULUCF has been the subject of two major reports by the Intergovernmental Panel on Climate Change. Additionally, land use is of critical importance for biodiversity.

Land-use change can be a factor in CO_2 atmospheric concentration, and is thus a contributor to climate change. IPCC estimates that land-use change contributes a net 1.6 ± 0.8 Gt carbon per year to the atmosphere. For comparison, the major source of CO_2, namely emissions from fossil fuel combustion and cement production amount to 6.3 ± 0.6 Gt carbon per year. This decision sets out the rules that govern how Kyoto Parties with emission reduction commitments account for changes in carbon stocks in land use, land-use change and forestry. It is mandatory for parties to account for changes in carbons stocks resulting from afforestation, reforestation and afforestation and voluntary to account for emissions from forest management, cropland management, grazing land management and revegetation.

The rules governing the treatment of land use, land-use change and forestry for the second commitment period are currently being renegotiated as part of the Bali Action Plan under the Ad-Hoc Working Group on Further Commitments for parties under the Kyoto Protocol. The most recent options for rule changes under consideration are summarized in a "Non-Paper" the co-chairs of the contact group on LULUCF.

The extent, and type of land use directly affects wildlife habitat and thereby impacts local and global biodiversity. Human alteration of landscapes from natural vegetation to any other use typically results in habitat loss, degradation, and fragmentation, all of which can have devastating effects on biodiversity. Land conversion is the single greatest cause of extinction of terrestrial species. An example of land conversion being a chief cause of the critically endangered status of a carnivore is the reduction in habitat for the African Wild Dog, Lycaon pictus. Of particular concern is deforestation, where logging or burning are followed by the conversion of the land to agriculture or other land uses. Even if some forests are left standing, the resulting fragmented landscape typically fails to support many species.

LAKE WINNIPESAUKEE ICE-OUT

Lake Winnipesaukee Ice-Out occurs when all the ice on Lake Winnipesaukee, the largest lake in New Hampshire, has broken up after winter. Over the years this has been decided upon by a variety of means; as of 2010, Dave Emerson makes the call. Emerson flies two to three times a day over Lake Winnipesaukee to check on the ice. Ice-Out is declared when the MS Mount Washington can make it to every one of its ports: Centre Harbor, Wolfeboro, Alton, Weirs Beach and Meredith. It is also considered the unofficial start to the boating season as well as the end of winter in New Hampshire. The earliest Ice-Out recorded was in 2010 on March 24, and the latest it occurred was in 1888 on May 12. Because the Ice-Out designation is decided upon by one man, it is highly unscientific. The call does not mean that the lake is entirely devoid of ice, nor does the MS Mount Washington actually go to each of its ports. It is simply believed that the ship could. Ice-Out records have been kept since 1887 as a way to keep track of when both commercial and passenger transportation lanes became usable in the lake. In 1974, Dr. William K. Widgert compiled data from known records to create a list of past dates.

Originally the Ice-Out designation was decided by people on shore. Later, for roughly 50 years, the call was made by Bob Aldrich, then, starting in the early 1980s, it was called by Alan Emerson. Since his death in 2002 it has been exclusively called by his son Dave. Ice-Out has occurred earlier on average in the last two decades of the 20th century than any of the preceding decades records have been kept, culminating in the ice-out for 2010 registering four days earlier than the 124-year-old record. This, as well as maple syrup season moving earlier over the same time span have led some to the conclusion global warming is involved.

KEELING CURVE

The Keeling Curve is a graph showing the variation in concentration of atmospheric carbon dioxide since 1958. It is based on continuous measurements taken at the Mauna Loa Observatory in Hawaii under the supervision of Charles David Keeling. Keeling's measurements showed the first significant evidence of rapidly increasing carbon dioxide levels in the atmosphere. Many scientists credit Keeling's graph with first bringing the world's attention to the effects that human activity was having on the Earth's atmosphere and climate. Charles David Keeling, of the Scripps Institution of Oceanography at UC San Diego, was the first person to make frequent regular measurements of the atmospheric carbon dioxide concentration, taking readings at the South Pole and in Hawaii from 1958 onwards. Prior to Keeling, the concentration of carbon dioxide in the atmosphere was thought to be affected by constant variability. Keeling had perfected the measurement techniques and observed strong diurnal behaviour with steady values of about 310 ppm in the afternoon at three locations. By measuring the ratio of two isotopes of carbon, Keeling attributed the diurnal change to respiration from local plants and soils, with afternoon values representative of the "free atmosphere". By 1960, Keeling and his group established the measurement record that was long enough to see not just the diurnal and seasonal variations, but also a year-on-year increase that roughly matched the amount of fossil fuels burned per year. In the substance, that made him famous, Keeling observed, t the South Pole the observed rate of increase is nearly that to be expected from the combustion of fossil fuel. Due to funding cuts in the mid-1960s, Keeling was forced to abandon continuous monitoring efforts at the South Pole, but he scraped together enough money to maintain operations at Mauna Loa, which have continued to the present day.

The measurements collected at Mauna Loa show a steady increase in mean atmospheric CO_2 concentration from about 315 parts per million by volume in 1958 to 385 ppmv as of June 2008. This increase in atmospheric CO_2 is considered to be largely due to the combustion of fossil fuels, and has been accelerating in recent years. Since carbon dioxide is a greenhouse gas, this has significant implications for global warming. Measurements of carbon dioxide concentration in ancient air bubbles trapped in polar ice cores show that mean atmospheric CO_2 concentration has historically been between 275 and 285 ppmv during the Holocene epoch but started rising sharply at the beginning of the nineteenth century. However, analyses of stomatal frequency in tree leaves indicate that mean atmospheric CO_2 concentration may have reached 320 ppmv during the Medieval Warm Period and 350 ppmv during the early Holocene.

Though Mauna Loa is not an active volcano, Keeling and collaborators made measurements on the incoming ocean breeze and above the thermal inversion layer to minimize local contamination from volcanic vents. In addition, the data is normalized to negate any influence from local contamination. Measurements at many other isolated sites have confirmed the long-term trend shown by the Keeling Curve, though no sites have a record as long as Mauna Loa. The Keeling Curve also shows a cyclic variation of about 5 ppmv in each year corresponding to the seasonal change in uptake of CO_2 by the world's land vegetation. Most of this vegetation is in the Northern hemisphere, since this is where most of the land is located. The level decreases from northern spring onwards as new plant growth takes carbon dioxide out of the atmosphere through photosynthesis and rises again in the northern fall as plants and leaves die off and decay to release the gas back into the atmosphere.

Due in part to the significance of Keeling's findings, the NOAA began monitoring CO_2 levels worldwide in the 1970s. Today, CO_2 levels are monitored at about 100 sites around the globe. Keeling died in 2005. Supervision of the measuring project was taken over by his son, Ralph Keeling, a climate science professor at the Scripps Institution.

INTEGRATED CARBON OBSERVATION SYSTEM

Integrated Carbon Observation System is a new strategic research infrastructure to quantify the greenhouse gas balance in Europe and adjacent regions. It consists of a harmonized network of ecosystem long-term observation sites, a network of atmospheric greenhouse gas concentration sites and a network of ocean observations. The network will be coordinated through a set of central facilities, including an atmospheric and an ecosystem thematic centre, a central data centre, an analytical laboratory and an oceanic thematic centre.

ICOS will provide the essential long-term observations required to understand the present state and predict future behaviour of the global carbon cycle and greenhouse gas emissions. It will monitor and assess the effectiveness of carbon sequestration and/or greenhouse gases emission reduction activities on global atmospheric composition levels, including attribution of sources and sinks by region and sector.

Standardized network will improved access to data and enable the development of flux products for research and political application. ICOS will be a state-of-the-art facility for the European research community. It will contribute the European share to global greenhouse gaz observations under GEO, WMO-GAW and GTOS programmes.

ICOS aims to build a network of standardized, long-term, high precision integrated monitoring of atmospheric greenhouse gas concentrations. The infrastructure will integrate terrestrial and atmospheric observations at various sites into a single, coherent, highly precise dataset. These data will allow a unique regional top-down assessment of fluxes from atmospheric data, and a bottom-up assessment from ecosystem measurements and fossil fuel inventories. Target is a daily mapping of sources and sinks at scales down to about 10 km, as a basis for understanding the exchange processes between the atmosphere, the terrestrial surface and the ocean. ICOS contributes to the implementation of the Integrated Global Carbon Observation System. The synergy between the atmospheric concentration measurements, the knowledge of local ecosystem fluxes on the other hand, has shown effective in reducing the uncertainties on carbon assessments. However, in Europe, observatories are all managed differently for each country and data is not homogenously processed. The value added impact of the infrastructure will allow an enhanced visibility and dissemination of European greenhouse gas data and products that are both long-term and carefully calibrated. ICOS seeks to meet the data needs of carbon cycle and climate researchers as well as those of politicians and the general public. ICOS will serve as the backbone to users engaged in developing data assimilation models of greenhouse gas sources and sinks, namely reverse modelling, which allows the deduction of surface carbon flux pattern.

A common data centre, the Carbon Portal put into place by ICOS, will provide free access to ICOS data, as well as to links with inventory data, and outreach material. This portal will allow the production web based tools for the survey of sources and sinks in near real time. ICOS will deliver the information in near real time with a quantification of the uncertainty associated with the results due to the use of several different models using different methodologies.

ICOS will enable Europe to be a key global player for in situ observations of greenhouse gases, data processing and user-friendly access to data products for validation of remote sensing products, scientific assessments, modelling and data assimilation. The implementation of ICOS will take place in two steps: During the Preparatory Phase starting in 2008 until 2011, the funding commitments will have to be endorsed by the governments and mother institutions, the building of the central facilities will be initiated, and the project will be technically developed up to the level of a demonstration year of full operation, but with a reduced number of observational sites.The ICOS preparatory phase includes 16 research laboratories from 12 European countries, coordinated in France. During the follow-up Operational Phase from 2012 until 2031, after the full scale deployment of the network, it will be run in an operational mode, and greenhouse gas concentrations and fluxes will be determined on a routine basis.

5

Land Use Patterns and Ecological Restoration

SIGNIFICANTLY CHANGE OF NATURAL LANDSCAPES

Human influence can significantly change natural landscapes through the alteration of natural ecosystem cycles, habitat loss and fragmentation. In many areas, industrial land use is the leading cause of habitat degradation, and contributes to rapid decline in biodiversity.

Over the past 70 years, human influence has resulted in significant changes in the forests of North America.

Effects of long-term climate change and short-term changes in land use and management persist within forests, and are known to influence processes of ecological succession. In human-dominated ecosystems, land use is a major determinant of the landscape structure, and changes in forests that result from land use are influenced by the social and economic factors controlling resource management.

By understanding the cumulative effects of past industrial management on a landscape, it is possible to better understand the ecological impacts of historical management practices (Axelsson and Östlund 2001).

If ecological restoration of landscapes altered by past land use is a management objective, the first priority is to document landscape decline and the need for restoration. Next, a successful ecological restoration programme should involve identifying the processes leading to landscape decline, to ensure control of the deleterious forces. If the processes are not controlled they may continue to cause degradation following ecological restoration. The most important role of land use history in ecological restoration is the identification of the anthropogenic forces acting on the ecosystem. In addition to understanding anthropogenic disturbances, it is also important to understand natural disturbance processes acting on the landscape, or processes which no longer play a key role.

UNDERSTANDING LAND

Understanding land use history in the context of ecological restoration of forested ecosystems is complicated by the fact that forests often superficially retain a "natural" appearance following disturbance caused by industrial management activities. In spite of their appearance, however, these systems may have undergone significant structural and/or ecological change. For example, the ponderosa pine (*Pinus ponderosa* Dougl. ex Laws.) ecosystems of the southwestern United States have undergone substantial changes resulting from the century-long practice of controlling fires in the interests of protecting timber and other resource values. Ponderosa pine forests are naturally regulated by fire, and the exclusion of fire has resulted in an increase in live and dead tree biomass, a decline in productivity of herbaceous plants, increasing susceptibility to insect and disease and an increase in crown fires. Creating an understanding of natural and anthropogenic disturbance history and vegetation change in these, and other, ecosystems sets the stage for more effective ecological studies, natural resource management and environmental modeling.

An additional and important role for historical studies in ecological restoration is in the characterization of the location, intensity and duration of natural disturbance patterns. In sub-boreal British Columbia, forest ecosystems have undergone significant changes since the initiation of industrial timber harvesting and forest management. Prior to the initiation of industrial land use, this region was heavily influenced by aboriginal management, and was infrequently disturbed by catastrophic natural events, such as fires (British Columbia Ministry of Forests and British Columbia Environment 1995). Understanding the natural disturbance regime under which the species, or ecosystem, in question evolved is key to evaluating the ecological consequences of forest management.

This study was conducted on the 13,032 ha JPRF, located approximately 40 km northwest of the village of Fort St. James, in the Sub-Boreal Spruce biogeoclimatic zone British Columbia, Canada.

Where coniferous species dominate in the SBS, the average mean fire interval was ca. 125 years. Forests in the SBS underwent frequent fires ranging in size from small spot fires to large fires that covered tens of thousands of hectares. It was typical for remnant patches of mature forest to remain standing in burned areas, creating an array of even aged stands ranging from a few to thousands of hectares with mature remnant patches interspersed. These mature remnant patches had an important function of increasing the structural complexity of the forest. In forests dominated by Douglas-fir the number and size of remnant patches increased due to the resilience of this species to wildfire, providing added structural diversity (British Columbia Ministry of Forests and British Columbia Environment 1995). In 1947 the dominant age cohort on the study area, the John Prince Research Forest (JPRF), was age class seven

(141-250 years) and it may be inferred that a large natural fire occurred in this ecosystem between 1807 and 1827 and the current forest was established following that disturbance

Unpublished Report

. The impacts of insect and disease were highly variable throughout the SBS. Also, wetlands, riparian areas and contiguous patches of mature forest created natural corridors throughout this region.

Following European settlement in British Columbia during the mid 1800s, forest ecosystems were influenced to a great extent by timber harvesting and livestock grazing. Simultaneously, many forest ecosystems were also indirectly altered through the suppression of wildfires. In much of central British Columbia, late successional tree species were harvested using selective harvesting and diameter limit logging practices. These practices have resulted in serious challenges for ecological restoration, conservation and forestland management.

Attempts to quantify landscape changes that result from anthropogenic causes are numerous; however, published accounts describing changes in forested ecosystems resulting from land use in sub-boreal British Columbia have focused on changes in landscape structure, and not the processes leading to these changes.

This chapter describes forms and sources of historical information relevant to understanding the land use history of forested landscapes in sub-boreal British Columbia. This information is used to describe the impact of: (1) harvesting effects, (2) mechanical innovation, and (3) silviculture trends on the JPRF in sub-boreal British Columbia.

FACTORS INFLUENCING LAND USE PATTERNS

Information from the oral histories and timber sale files were correlated, and combined with unpublished data and reports providing additional information on land use history. The data was compiled into a composite land use history for the study area.

Integrating government documents and reports with oral histories creates a comprehensive picture of the industrial management practices. Cultural practices have significantly changed the vegetation on the JPRF from the 1940s onward. An understanding of these practices, and the resulting changes on the landscape structure, is critical for developing indicators for ecological restoration. An indicator is a quantitative or qualitative variable which can be measured or described periodically to monitor changes. In this context indicators refer to vegetation composition and configuration on the landscape, and the extent to which change has occurred based on industrial land use. Three factors dictate industrial land use patterns and the degree of anthropogenic landscape change on the JPRF, including: (1) harvesting effects, (2) mechanical innovation, and (3) silviculture trends.

INFORMATION SOURCES AVAILABLE

Historic information sources, in the form of oral histories and government reports, were used to analyze trends in land use from 1940 to 1970 on the JPRF. This includes the period between the initiation of industrial logging activities and the availability of modern records. Oral histories were conducted with local knowledge holders who lived or worked on the JPRF between 1940 and 1975, and the interview transcripts were analyzed for words, phrases or themes reflecting information related to past industrial land use practices. Interviewees were selected to represent a diversity of perspectives related to industrial land management activities. In addition, a collection of interview transcripts completed with members of the Tl'azt'en community and other resource professionals were reviewed to obtain related information. Oral histories were used to corroborate information found in government reports, and helped to clarify information in records and related to conventions used in the past. Timber sale files were also obtained and analyzed for the study area. Timber sale files are a form of government report that documents past timber harvesting activities in British Columbia. Timber sale files were examined to characterize the distribution of management activities across the JPRF between 1940 and 1970 to determine ownership patterns across the landscape, and to understand changes over time in management, rules and standards for timber harvesting. In British Columbia, timber sale files remain the most consistent and accessible record of land use history in forests.

Harvesting Effects in Forest Service

The British Columbia Forest Service was established in 1912 and given the responsibility of protecting and allocating timber resources within British Columbia. Between the early 1900s and the mid 1970s, area-based sales were typically used to assign timber-harvesting rights. Companies or individuals were responsible for locating merchantable stands of timber, and after a suitable stand had been selected a written request to obtain the area for timber harvesting purposes was submitted. If approved, the Forest Service would allocate the rights to harvest the timber to the applicant. In the context of the JPRF, three factors were dominant in determining the likelihood harvesting activities having occurred, including: (a) historical market demand, (b) proximity to waterways, and (c) terrain.

IMPORTANT FACTOR OF HISTORICAL MARKET DEMAND

Historical market demand was an important factor affecting trends in harvesting activities on the JPRF. Two dominant sources of market demand drove harvesting in this region: processing activities related to the development of cinnabar deposits on Pinchi Lake, and lumbering operations on Tezzeron Lake.

Development of Cinnabar Deposits at Pinchi Lake - The Pinchi Mercury Mine is located adjacent to the southern portion of the JPRF on Pinchi Lake. The Consolidated Mining and Smelting Corporation of Canada (Cominco) developed cinnabar deposits located on the north side of Pinchi Lake. Cinnabar, when harvested, releases elemental mercury which was used to construct fuses, batteries and precision instruments during World War II. Cordwood was harvested from stands east of the mine, and used to fuel a crusher, kiln and condenser for the production mercury from the ore. In 1944, the Allies re-secured supplies of mercury in Spain and Italy, and the Pinchi Mercury Mine was subsequently closed, and later re-opened in the 1960s: lumbering operations, however, continued in the area into the early 1950s.

Lumbering operations at Pinchi Lake in the 1940s focused on harvesting Douglas-fir trees to be used in building permanent structures associated with mine operations. In addition, a substantial amount of cordwood was harvested on the south side of the JPRF as fuel for the mine. All available tree species were harvested for cordwood, including Douglas-fir, paper birch and trembling aspen (*Populus tremuloides* Michx.); however, operators showed a preference for Douglas-fir and paper birch. Paper birch was the preferred species because it gave off intense heat, but it was reported as being relatively uncommon in the area and Douglas-fir was the most readily available species that emitted substantial heat, and was therefore the predominant species harvested for fuel and lumber to support the mine.

Lumbering Operations on Tezzeron Lake - During the 20 years following World War II, Douglas-fir and interior spruce were the preferred commercial species in the area, and harvesting in the JPRF was focused on these species. As the cost to haul lumber was high in remote areas, such as that now occupied by the JPRF, companies sought out prime stands of timber to cover the high costs associated with transportation. Records indicate the Douglas-fir forest cover in the northern part of the JPRF near Tezzeron Lake was of a quality uncommon in central British Columbia, and considerable harvesting occurred to extract Douglas-fir and interior spruce from that area. Patches of lodgepole pine and subalpine fir were present across the landscape, and although they were merchantable, they were generally avoided in harvesting operations during this period due to the low commercial value of these species at the time.

PROXIMITY TO WATERWAYS

Historically, water was the most readily available means of transporting timber and, as a result, industrial activities developed in close proximity to rivers, lakes and oceans. The position of the JPRF on an isthmus between two lakes significantly affected harvesting patterns, since water was used as the primary means of transporting logs from timber sales to the sawmills. Trees were harvested on the lakeshores and transported east by water to the permanent and portable mills. Close proximity to such large bodies of water increased harvesting activities on the north and south sides of the JPRF. This trend is reflected both in the location of permanent mill sites and in the location of timber sales on the JPRF. Landings were built on the shore of a lake or river near an active timber sale. Felled trees were hauled to landings, where they were cut into smaller lengths. Operators often cut logs on the ice during the winter months and left wood on the ice until the breakup that occurred as the temperature warmed. When the lake ice melted in spring, logs would be floated on the water, and the sawdust and other debris would settle at the bottom of the lake. Throughout the spring and summer months, logs were secured with a barrier composed of a chain of floating logs enclosing other free-floating logs (*i.e.* a boom) which was moved to the sawmill with a tugboat or river barge. Permanent mills on the JPRF were built on lakes near the best and most accessible stands of timber. In cases where timber sales were located a considerable distance from the sawmill, damage inflicted to the trees from dragging trees during skidding was substantial. This damage impacted the value and merchantability of the trees. Operators, therefore, turned to water as the primary means of transporting logs over long distances.

CHARACTERIZED BY STEEP TERRAIN

Many areas of the JPRF are characterized by steep terrain. Equipment limitations made logging difficult in the steep and complex areas, and crawler tractors were used to skid timber down very steep grades. For example, areas like Pinchi Mountain (in the centre of the JPRF) were not impacted heavily by timber harvesting in part because of the difficult terrain and distance from the water.

LANDSCAPE OF MECHANICAL INNOVATION

In this context, mechanical innovation refers to the impacts on the landscape from different forms of equipment used for timber harvesting. As advances in mechanical equipment ensued, these changes were reflected on the landscape of the JPRF. Three factors which reflect changes in mechanical technology and the means by which the landscape is changed, including: (a) advances in sawmill equipment, (b) advances in wood transportation, and (c) road construction technology.

ADVANCES IN SAWMILL TECHNOLOGY

There were both portable and permanent sawmills on the JPRF. Sawmills responded to market definitions of premium tree species and sizes, and trees were harvested accordingly. Throughout the 1950 and 1960s, there were approximately 20 to 25 portable sawmills in the forests north of Fort St. James at any given time, many of which were situated between Pinchi and Tezzeron Lakes. These portable mills were small operations, dispersed throughout the JPRF, producing approximately 18,000 to 20,000 BFT (62.6 to 69.6 CBM) of lumber daily. The mills were moved from one timber sale to the next, with scrap wood and other waste from the mill typically left behind. Remnants of sawmill operations may be found on the landscape using sawdust piles as evidence of their past location[7]. Portable mills were generally established in heavily forested areas away from significant waterways. Skidding logs to these mills was limited by technology and economics, and was typically restricted to a distance of ¼ mile (0.4 km). As a result, portable mills were typically dismantled and moved to a new timber sale each season, and reassembled again at the new site.

There were three known permanent sawmills on the JPRF. The first mill was located on Pinchi Lake near the Pinchi Mercury Mine. This was a relatively small operation, held roughly four small timber sales at one time, and employed approximately 20 individuals. The second permanent sawmill, Cassiar Forest Products, was erected on Tezzeron Lake near Sunshine Island in 1955. This mill operated on Tezzeron Lake for nine years, and produced dimensional lumber until 1964, when it was put out of business by competition from larger milling operations in the area. Throughout the nine-year operation, the mill expanded from a capacity of five employees to 37. The year following the closure of the mill, a resort was established on the old mill site which operated until 2002 when it was subsequently bought out by the JPRF.

The third permanent mill on the JPRF was owned and operated by Park Brothers Ltd., and ran for fourteen years between 1955 to 1969. This mill ran a double shift producing approximately 50,000 to 60,000 BFM per day (174 to 209 CBM), and was considered the most stable lumbering operation on the JPRF. Trees were cut to size and length at the mill on Tezzeron Lake, and logs were shipped to a planing mill approximately 100 km away to be sorted, have the bark peeled, and cut into lumber. The two larger operations focused harvesting activities on the northern side of the JPRF and focused on harvesting Douglas-fir and interior spruce.

Mills on the JPRF grew in size to increase the production of lumber and compete with larger mills in the area. Several mills avoided cutting large trees, as equipment at the mills were unable to handle large logs. The British Columbia Forest Service did not want companies to leave large trees on timber sales; so much of the wood from large trees would often go to waste as trees would be cut but not processed. Cassiar Forest Products was the first mill on Tezzeron Lake to close, followed by Park Brothers Ltd. four years later. Park Brothers Ltd. was taken over by Takla Logging in 1969, which was later called Canadian Forest Products Ltd. (Canfor). After Takla Logging began controlling operations in the late 1960s, processing operations were centralized in the town of Fort St. James.

Road Construction Technology

Prior to harvesting and mining activities the JPRF was very isolated, with no evidence of roads. Aboriginal users describe an intricate trail system in the area that facilitated foot transportation from settlement areas to the *keyohs*, or family hunting and trapping grounds. The first evidence of road building in the JPRF came with the establishment of the Pinchi Mercury Mine. Road development increased in the southern portion of the JPRF with mine expansion and the establishment of several sawmill operations. The first road to the mill on Tezzeron Lake was constructed in 1951, and Park Brothers Ltd. built a second road to Tezzeron Lake in 1956. Park Brothers Ltd. built most of the road networks on the JPRF between 1955 and 1969 to ease accessibility to timber sales.

The majority of the roads that existed prior to 1970 on the JPRF were winter roads. It was difficult to build summer roads in many areas, due to inadequate equipment and the wet terrain common in the valley bottoms, where road building was most economical. When roads were built in wet areas, the builders used non-merchantable deciduous species to stabilize wet roads (corduroy roads) in order to make roads passable and more durable. When harvesting in areas that were characterized by difficult terrain and steep grades, roads were often built through creek bottoms. When Takla Logging took over operations on the JPRF in 1969, they upgraded several winter roads to create permanent roads, some of which still exist today. Other roads that were not upgraded by have become inaccessible or inactive.

Advances in Wood Transportation

Harvesting methods on the JPRF varied temporally and spatially across the landscape. Harvesting typically occurred during the winter, and after the trees were felled the logs were skidded to haul roads with a team of horses and loaded onto sleighs. Ice roads were built to facilitate pulling the sleighs, and significant attention was placed on creating the most efficient logging operations using horses and sleighs. Mechanical innovation, and the introduction of the crawler tractor, led to the expansion of lumbering operations on the JPRF. Crawler tractors were used for a diverse range of tasks, including replacing horses as the dominant log skidding technology. A winch was fastened to the crawler tractor, and logs were simply dragged behind the machine. The increased power offered by the crawler tractor allowed operators to skid logs up to ¼ mile (0.4 km) and haul larger loads, often more than one log at a time (depending on the size of the trees). Wagons and sleighs, formerly drawn by horses, were adapted for use with crawler tractors, and the addition of a blade to the front of the crawler tractor provided a simple means of facilitating road construction. The next major change for harvesting operations on the JPRF was the introduction of the arch truck. The arch was simply a device mounted on the back of a truck to elevate one end of the logs during skidding. Like the crawler tractor, the arch truck allowed operators to move logs over longer distances and handle much heavier loads. These two systems, the crawler tractor and the arch truck, facilitated the expansion of harvesting activities into previously inaccessible areas, and increased the volume of wood harvested.

EFFECTS OF DIFFERENT SILVICULTURAL PRACTICES

The effects of different silvicultural practices are a significant factor in change in landscape condition on the JPRF. Historical timber harvesting activities were not undertaken with the same intent as silviculture is typically applied today. The term silviculture refers to the "treatments applied to forests to maintain and enhance their utility for any purpose". Historic silviculture practices on the JPRF were primarily focused on short-term profit and job creation. Silvicultural practices during this period were experimental and were essentially based on trial and error. On the southern portion of the JPRF, where timber was harvested to fuel the mine, small clearcuts tended to dominate the landscape. Where the forest was selectively logged for Douglas-fir and interior spruce near Tezzeron Lake, the remaining stands are distinct in character.

In areas where selective harvesting prevailed, utilization standards for tree species and size underwent many changes from the 1940s to the 1970s, and records of past harvesting operations on the JPRF reflect these changes. Single-tree selection was the predominant silvicultural method used in the 1940s, and this practice changed to diameter-limit harvesting in the years that followed.

Diameter-limit harvesting was the dominant silvicultural practice documented on the northern portion of the JPRF. Diameter-limit harvesting involves removing the highest quality and most vigorous trees greater than a set diameter from a site, leaving the suppressed understory and poorer quality trees to naturally restock the site if conditions were favorable. Diameter-limit harvesting practices typically do not consider future stand conditions, and alter forest growth patterns of regeneration, and succession. During the 1950s and 1960s, operators were not permitted to harvest trees less than 140 years of age. The diameter limit established by the Forest Service was 18 inches, thus trees smaller than 18 inches (46 cm) in diameter were not harvested. The diameter limit was later reduced to 16 inches (41 cm), indicative of the increasing scarcity of large trees and changes in technology and markets. As time progressed, the diameter limit continued to shrink, and operators were permitted to harvest progressively smaller trees. Multiple waves of harvesting occurred in the northern portion of the JPRF, as harvesting rules changed and the diameter limit was reduced, making previously unmarketable trees within the stands merchantable. For example, in the mid 1950s strip cutting was introduced, where harvesting was done in strips and "leave strips" were left behind to naturally regenerate the forest. Most often, operators attempted to include Douglas-fir and interior spruce in harvested strips and leave lodgepole pine and subalpine fir in the leave strips. These practices may have significant impacts on current forest conditions. During the 1950s and 1960s, Douglas-fir was the most commercially valuable species throughout central British Columbia. This led to significant impacts on the Douglas-fir component of the forest, altering the representation of Douglas-fir in these forests today.

Following the 1970s, clearcutting replaced diameter-limit harvesting as the dominant silviculture system. In some areas, patches that were previously cut using diameter-limit harvesting were revisited and trees that were not harvested originally were then cut.

RESTORATION OR MANAGEMENT OF FOREST ECOSYSTEMS

Research focused on restoration or management of forest ecosystems must consider the historical context in which past communities developed, and the subsequent changes brought about by industrial land management. This study marks the first in-depth attempt to understand the historical industrial land use patterns that have impacted landscape structure in sub-boreal British Columbia. The oral sources obtained for use in this study were instrumental in reconstructing past land use patterns on the JPRF. Through this work the three factors that have dictated land use patterns on the JPRF have been identified, including: (1) harvesting effects, (2) mechanical innovation, and (3) silviculture trends. All of these factors have lead to contemporary changes apparent on the JPRF, and have created the primary disturbance regime.

Reference periods vary for different landscapes, depending on the natural rotation of the forest, and the time period when industrial land use activities occurred.

The early 1940s were selected as a suitable point in the natural forest rotation to study, as industrial management was the primary factor leading to observed degradation on the JPRF and fire suppression practices were in place for several years at that time. Also, information sources were available for the 1940s, but not before. Areas that were not harvested or managed during this time were left to develop naturally. These stands, however, were often small remnant patches of the original forest. Such stands are often were heavily impacted by edge effects. On landscapes such as the JPRF, unharvested areas often occur on less productive sites, or sites that have limited access. Understanding where harvesting activities dominated the landscape is important in establishing target areas for ecological restoration, but it is also important to realize that those areas not directly impacted by harvesting activities may not be generally representative of the original forest.

The JPRF is located within Tl'azt'en traditional territory, and Tl'azt'en Nation managed the flora and fauna of their territory for subsistence purposes prior to 1940. Traditional management occurred concurrently with industrial development in the years following 1940. Tl'azt'en people managed the land primarily for harvesting salmon, whitefish, char, berries, and small birds and mammals. Tl'azt'en Nation organized their territory into *keyohs*, or family use areas, and five permanent village sites were located within Tl'azt'en territory: however, there is no evidence that permanent village sites existed within the boundaries of the JPRF.

Trapping was concentrated in riparian areas, old roads and trails, and focused on small mammals. Hunting took place through the forest and focused on large game. In some areas (including the Pinchi Lake area), there have been reports that aboriginals traditionally burned wild hay meadows for horse and cattle grazing. The impacts of aboriginal management have often been understated and we tend to generalize these activities despite the site-specific impacts on the land.

A question that needs to be answered to better understand the development of landscape structure is the role of traditional land management on the JPRF. This study has focused on industrial land management on the JPRF that began in the 1940s; however, cultural practices of aboriginal people are known to have altered ecosystems prior to this time. This information is of great importance, as ecological restoration must not only reintroduce the structure of the historic ecosystem and eliminate industrial land practices that were the primary cause in landscape change, but also reproduce the forces (natural and anthropogenic disturbance) that shaped the target community. A more comprehensive understanding of traditional aboriginal land management practices on the JPRF is critical to understanding forces which shaped the historic ecosystem.

Industrial land use in forests of sub-boreal British Columbia has resulted in changes in both the composition and configuration of the landscape. Landscape composition refers to the variety and abundance of patch types on a landscape, while landscape configuration refers to the spatial distribution of these patch types. Specifically, mechanical innovation impacts both landscape composition and configuration, as technological advances influence the species and size of trees harvested (composition), and increase the extent of areas where it was possible to harvest timber by facilitating movement into previously inaccessible areas (configuration). Factors affecting the likelihood of a stand being harvested also impact landscape composition and configuration. Historic market conditions dictated the species and size of trees harvested (composition), and proximity to water and terrain dictated which stands were harvested on the landscape (configuration). Finally, the silviculture systems employed also impact both composition and configuration, as changes in silviculture practices primarily determined changes in size of trees harvested and left behind (composition). Different silviculture systems also leave different patterns on the landscape (configuration).

This chapter demonstrates how land use patterns have impacted landscape structure on the JPRF, and understanding how forests have been influenced by past land use is essential to successful management of forests. Accordingly, it is critical to understand if these changes that have come about as a result of industrial land use are within the natural range of variation for this ecosystem. The next step in this research involves quantitatively analyzing changes in landscape condition that have resulted from industrial land management, to further assist in defining goals for ecological restoration on the JPRF.

LAND USE ON LANDSCAPE STRUCTURE

The rate and direction of changes resulting from industrial land use in forests over the past century may be outside the natural range of variability, and these changes may result in habitat loss and fragmentation. In contrast to natural disturbances in forests, industrial land use tends to create simplified patches, causing a reduction in patch size, an increase in patch isolation, and more edges. As changes in landscape structure have direct impacts on ecological function, it is important to maintain large tracts of the landscape with representative seral stages, species distribution, patch sizes and corridors if natural ecosystem functioning is to be sustained.

This may be accomplished in part through forest management practices that are sustainable, and through the allocation of parks and protected areas. Although there is some uncertainty, it has been shown that changes in landscape pattern resulting from industrial land use may lead to the decline of many forest-dependant species. Using natural, or at least historical, forest landscape structure (*i.e.* forest composition and configuration) to design management regimes and to guide ecological restoration may assist in maintaining natural ecological functioning, and therefore maintain natural levels of biodiversity.

In sub-boreal British Columbia, ecological responses to changes in landscape structure resulting from industrial land use are largely unstudied. Available studies of forest management impacts over large areas of the interior of British Columbia indicate these forests are in the early stages of fragmentation. Studies of the impacts of forest management at larger spatial scales (*i.e.* over smaller areas) are limited. It is widely accepted that ecological restoration of landscapes impacted by industrial land use and other anthropogenic disturbances requires an understanding of the changes that have resulted from the cumulative impacts of land use. It is important, therefore, to further understand the extent of changes in landscape condition resulting from industrial land use in sub-boreal British Columbia, and to evaluate the need for, and relevance of, ecological restoration in these forests.

IMPORTANT CONTRIBUTOR TO THE ECONOMY, AND FORESTRY

In British Columbia, natural resources are an important contributor to the economy, and forestry is the predominant sector. As a result, timber harvesting and related road construction constitute the dominant form of development activities in forests. In the relatively recent past, forest resources in British Columbia were considered abundant, and as a result the forests were harvested with a limited vision for the future forest condition. Prior to 1970, timber leases were short-term, and replanting following selective harvesting was not common. Harvested stands were left to naturally regenerate, and as a consequence these forests retained little of the natural landscape condition. For example, the past half-century of forest management in sub-boreal British Columbia has focused on harvesting Douglas-fir. This species reaches the northern extent of its natural range in the interior of British Columbia, and concern has been expressed regarding changes in the representation of Douglas-fir in the forests. Selectively cut Douglas-fir stands often become dominated by more shade-tolerant understory species, such as subalpine fir, while clearcut Douglas-fir stands are often replanted with lodgepole pine or interior spruce. Understanding the extent to which the characteristics and distribution of Douglas-fir and other tree species has changed as a result of forest management and other industrial land use activities is important for guiding ecological restoration in sub-boreal British Columbia.

Road development is another form of development that occurs in forests along with timber harvesting, and the presence of extensive road networks may alter natural ecosystem functioning (Forman and Alexander 1998, Trombulak and Frissell 2000). These impacts may have more significant ecological implications than the habitat loss that results from timber harvesting, as roads create high contrast edges and cause fragmentation. Natural processes such as soil erosion, sedimentation and landslides may be increased with the presence of roads, and roads may change natural disturbance cycles. By enhancing access into forests, roads increase the frequency of fires caused by humans and increase hunting activities. Roads also increase the movement of exotic species through otherwise remote areas, and create a barrier against species dispersal. In some forests, roads have been shown to have a more lasting impact on landscape structure than timber harvesting.

A practical method for quantifying these and other changes in landscape condition in forests is through the analysis of historical records. Changes in landscape structure can be measured directly using permanent sample plots, however these plots must have been established during the past and require continued monitoring over time. Aerial photography, topographic and planimetric maps are historical records that may also be used to illustrate changes in landscape structure over time. Improved aerial photography technology for mapping, inventorying and monitoring vegetation and landscape features, combined with progress in the production of orthorectified aerial photographs in digital form, make it possible to monitor changes in landscape structure over long temporal and large spatial scales. Until recently, the analysis of these historical sources, and black and white aerial photography in particular, has been limited to visual interpretation. The introduction of digital image processing and geographic information systems (GIS), however, allow for the detection and quantitative analysis of changes in landscape structure within ecosystems and along ecosystem boundaries.

Criteria and indicators provide framework for developing forest management strategies, and serve as a tool to monitor progress towards sustainable forest management. A *criterion* is a category of conditions or processes by which sustainable forest management may be assessed, and is characterized by a set of related *indicators.* The latter are monitored periodically to assess change. An indicator is a quantitative or qualitative variable and can be measured or described which can be observed periodically to monitor trends. In this context, landscape indices may be used as criteria and/or indicators for ecological restoration, and to measure changes in landscape structure and function that have been altered by industrial land use practices.

Landscape indices are response variables that may be used to quantify landscape structure and indirectly measure ecological processes. These indices are especially useful for interpreting spatial changes in landscape structure over time making certain, limited, predictions about ecological processes (Schumaker 1996, Giles and Train 1999, Tischendorf 2001). For example, it has been previously demonstrated that many indices are redundant or ambiguous.

It is generally accepted that a small, limited set of landscape indices offers the most insight into both landscape patterns and ecological processes. Spatial evidence of landscape structure from historical records such as aerial photography will assist forest managers assess natural and/or human induced changes in landscapes over time. The purpose of this study is to use historic aerial photography with image processing and GIS as a technique to quantify changes in landscape structure on the John Prince Research Forest (JPRF) between 1947 and 1999. The intent of this chapter is to provide insight into changes in landscape structure that have incurred in selectively harvested forests of sub-boreal British Columbia and to characterize changes at the landscape level which may have implications for ecological functioning.

DATA PREPARATION AND DEVELOPMENT

Contemporary Forest Conditions (1999) - Two sources of data were used to describe contemporary forest conditions on the JPRF. The first source was the British Columbia Ministry of Forests forest cover inventory database, which contains spatial data on forest cover and associated attributes. This was used to assess the extent of timber resources. The second source was Terrain Resource Information Mapping data (TRIM II), obtained through the Geographic Data British Columbia Branch of the Ministry of Environment, Lands and Parks. TRIM II provides current topographic data, cultural features, land cover, surface features, transportation features (roads, railways, pipelines, bridges) and water features (lakes, rivers, wetlands) at a scale of 1:20,000.

Historical Forest Conditions (1947) - Historical forest conditions were characterized using black and white aerial photography. British Columbia has been extensively covered with aerial photography since the 1920s, providing a source of historical data for most of the province. Panchromatic aerial photography from 1947 was obtained from the JPRF, and the photographs were converted into digital images. Ground control points from TRIM II data were used to rectify photographs to the Universal Transverse Mercator (UTM) map projection, and the images were further geometrically corrected using a digital elevation model to create an orthophotograph. An orthophotograph is a rectified photograph that presents objects in true planimetric arrangements, where distortions due to tilt, variations in height and topography are removed. Orthophotographs are geometrically equivalent to conventional line maps (or TRIM II) and can be used to analyze distances, angles, positions and areas. The orthophotograph was converted to digital format and timber-typed using current forest inventory standards. The resulting database was converted to GIS maps for analysis. Polygons were delineated based on species composition, age class, height class, stocking class and crown closure. Additional landscape features such as lakes, swamps, streams, landslides, rock, clearcuts, roads, non-commercial or non-productive brush, and clearings were also defined. Industrial Forestry Service Ltd. (Prince George, British Columbia) converted the photographs into digital format and performed the timber typing.

DATA STRATIFICATION

The historical and contemporary data was stratified for this analysis using three schemes: (1) vegetation cover type, (2) seral stage, and (3) a combination of vegetation cover type and seral stage. The vegetation cover type map was stratified based on leading species (*i.e.* the dominant tree species in the stand). Leading species was defined as the species with the greatest abundance on a forest inventory label by species composition. The "brush" stratum includes polygons classified as non-productive or non-commercial brush. The seral stage map was created by reclassifying inventory age classes into four seral stages, corresponding in part to seral stage definitions for Natural Disturbance Type 3 (within which the JPRF falls) in the Biodiversity Guidebook (British Columbia Forest Service and British Columbia Environment 1995). The following age ranges were used: early (<40 years), mid (40-99 years), mature (>100 years), and old (>140 years). Although the mid seral stage is not defined in the Biodiversity Guidebook (1995), it was included here. The combined vegetation cover and seral stage distribution map included 26 strata, with all deciduous species – black cottonwood (*Populus trichocarpa* Torr. & Gray), paper birch and trembling aspen - combined into one vegetation type. The maps were converted to raster format with a cell size of 25 m for analysis.

LANDSCAPE STRUCTURE AND DATA ANALYSIS

Landscape structure (composition and configuration) was characterized for the six aforementioned maps using FRAGSTATS and the results from 1947 and 1999 were compared. A set of landscape indices were used to quantify elements of landscape structure based on the assumption that limited predictions regarding ecological processes (*i.e.* response variables) can be made by analyzing landscape structure (Forman and Gordon 1986, Tischendorf 2001). In this context, criteria used to assess sustainable forest management with respect to ecological restoration include landscape composition and landscape configuration. Landscape composition refers to the area, in percent or hectares, of the landscape analysis unit comprised of a particular stratum, and is typically used to estimate the extent of available habitat and by inference the degree of biodiversity. Landscape configuration refers to the arrangement of these strata across the landscape, and can be interpreted as indices of fragmentation (Forman and Godron 1986, McGarigal and Marks 1995). Area by stratum was used as an indicator of landscape composition. Specifically, class area (ha) tabulates the amount of the landscape composed of each stratum. Habitat loss is a fundamental concern with landscape change, and it is important to understand the amount of habitat (to the limits of what can be inferred from the vegetation type and seral stage) that exists on the landscape. Percentage of the landscape occupied by each stratum was computed to quantify area in relative terms as percent of total landscape area.

The complexity of patch shapes as compared to a standardized shape (*i.e.* a circle) was analyzed using mean shape index. Patch shape influences several ecological processes, however the most important influence appears to be related to edge effects. The edge of a patch can have a significantly different environment than the patch interior, resulting in a different array of species. Different organisms show variable responses to changes in the edge of a patch. For example, the American marten (*Martes americana*), an interior forest species, demonstrates a negative response to an increase in edge, where other species such as white-tailed deer (*Odocoileus virginianus*) thrive when edge habitat is increased. The co-variation of Euclidean nearest neighbor (ENN) distance refers to the distance between two patches of the same cover type based on an edge-to-edge comparison. The most significant interpretation co-variation of ENN relates to patch isolation and patch connectivity. It has been demonstrated that an increase in patch isolation, demonstrated by an increase in co-variation of ENN, explains why fragmented landscapes contain fewer bird species that contiguous habitats.

As road development that occurred in conjunction with timber harvesting may have significant (and independent) ecological impacts, a preliminary analysis of road development on the JPRF was conducted. To access changes in road development resulting from industrial land use between 1947 and 1999, road length (km) was measured and compared between the two time periods.

CHANGES IN VEGETATION COVER TYPE

Landscape Composition - There have been changes in vegetation cover type on the JPRF between 1947 and 1999. Landscape composition was measured using class (or strata) area, and percentage of the landscape occupied by that stratum. On the JPRF, there have been decreases in the representation of strata that were harvested for timber. For example, the representation of lodgepole pine has decreased (34% to 14%), and the representation of Douglas-fir has decreased (28% to 24%) and the Douglas-fir component has shifted from the northern portion of the JPRF to the interior. The relatively small decrease in the extent of Douglas-fir, which was heavily impacted by harvesting and often not replanted or regenerated, may be an artifact of natural successional processes. Mixed lodgepole pine/Douglas-fir stands have matured since 1947, and the longer-lived Douglas-fir has begun to dominate these stands as the shorter-lived lodgepole pine has become overmature and subsequently died out. This may also partially explain the precipitous drop in the extent of lodgepole pine dominated stands. Also, the decrease in lodgepole pine may be related to fire suppression, and the difficulty naturally regenerating lodgepole pine in the absence of fire.

Although interior spruce was harvested for timber, there has been an increase in the composition of that species (20% to 32%) likely due to replanting of interior spruce. The increase in subalpine fir (0.8% to 9%) may reflect the release of this suppressed understory species when Douglas-fir and interior spruce overstory were removed through harvesting practices. There has been an increase in area classified as brush (1% to 2%), which may have been created by selective harvesting when merchantable species were removed (Douglas-fir, interior spruce and a lesser extent lodgepole pine and subalpine fir) and other species were left to naturally regenerate. There have not been significant changes in the composition of those strata that which were not harvested for timber, including black spruce, black cottonwood, paper birch, trembling aspen, and lakes and wetlands.

Landscape Configuration – There has been an increase in fragmentation in strata where timber harvesting occurred, indicated by an increase in patch density and a decrease in mean patch size. Douglas-fir and lodgepole pine strata became increasingly fragmented between 1947 and 1999. Although there is only a relatively small area occupied by subalpine fir and paper birch on the JPRF, these stand types also demonstrated an increase in fragmentation. Changes in interior spruce are ambiguous as this stratum shows an increase in patch density and a decrease in patch size indicating a move towards increased fragmentation. This contradicts observed patterns on the JPRF where the number of interior spruce leading stands in larger patches has increased resulting from reforestation practices. Black cottonwood shows a decrease in patch density and an increase in patch size, but the small overall representation of this stratum on the landscape makes interpretation difficult. It is possible that changes in black cottonwood is a result of the subjective nature of the inventory methods used, rather than a real change in the stratum.

All vegetation cover types (except black spruce) have undergone a decrease in patch complexity, based on the mean shape index. The decrease in patch complexity indicates a decrease in available edge habitat, and a movement towards less complex patches in harvested stands. This change has different ecological implications depending on the species in question. Also, vegetation types heavily impacted by timber harvesting have become less connected. For example, in Douglas-fir and lodgepole pine leading stands there has been an increase in patch isolation as measured by the co-variation of ENN. Also, there has been an increase in patch isolation in subalpine fir and paper birch leading stands as measured by the co-variation of ENN, despite the overall increase in the representation of these stands across the landscape. As there was not significant enough area occupied by subalpine fir and paper birch leading stands in 1947, there was no value recorded for co-variation of ENN at that time.

In addition to changes in vegetative strata, there has been a change in the lake and wetland strata where patch density has increased and mean patch size has decreased, indicating increased fragmentation. Also, the increase in co-variation of ENN indicates a decrease in connectivity in this stratum. As the changes in area classified as lake and wetlands are minute (343 ha to 349 ha), the changes reported here are not believed to be significant. There have been changes in the strata harvested using clearcut silviculture systems between 1947 and 1999. Because there is no data on the areas selectively harvested it is difficult to make assumptions about this stratum.

CHANGES IN SERAL STAGE DISTRIBUTION

Landscape composition – The seral stage distribution on the JPRF has changed between 1947 and 1999. There has been a decline in seral stages subjected to timber harvesting on the JPRF, including mature seral stages (63% to 48%) and mid seral stages (20% to 16%). Natural and artificial regeneration in harvested stands has lead to an increase in the representation of early seral stages (14% to 31%). This increase in early seral stages does not appear to be of the magnitude observed when the mature cohort that dominates the forests was established (*i.e.* the increase in early seral stages is not of a magnitude that would indicate the beginning of a new rotation). Old seral stages have also increased (0.7% to 33%) and this may be the fire suppression policies combined with the ageing of stands in unharvested areas.

Landscape Configuration – There has been a dramatic increase in fragmentation of old-seral stages as reflected by an increase in patch density. Also, mature-seral stages have become increasingly fragmented as shown by an increase in patch density and a decrease in mean patch size. Early seral stages have become considerably less fragmented as reflected by a decrease in patch density and an increase in patch size. Changes in mid seral stages were not substantial.

As a result of timber harvesting, mid, mature and old seral stages have undergone a decrease in patch complexity as measured by a decrease in mean shape index. In early seral stages mean shape index has increased resulting in an increase in patch complexity, and therefore in edge habitat. Ecological consequences of this change may have positive impacts on species such as white-tailed deer, which thrive in edge habitat and young forests.

In strata heavily impacted by timber harvesting there has been a considerable decrease in patch connectivity. For example, co-variation of ENN has increased in mid, mature and old seral stages, indicating an increase in patch isolation and thus a decrease in connectivity.

Changes in Vegetation Type

Landscape Composition - Changes in seral stage representation by vegetation type can be characterized by combining leading species and seral stage. As Douglas-fir, interior spruce and lodgepole pine leading stands constitute the greatest proportion of the JPRF in 1947, the discussion is focused on these cover types.

Commercially valuable conifers have undergone a decline in mature seral stages as a result of timber harvesting in these age classes. For example, there has been a decrease in the representation of mature Douglas-fir (22% to 21%), a decrease in the representation of mature interior spruce (17% to 16%), and a decrease in the representation of mature lodgepole pine (20% to 8%). As well, all commercially valuable conifers have shown an increase in old seral habitat essentially showing this forest becoming older. For example, there has been an increase in old Douglas-fir (0.7% to 15%), an increase in old interior spruce (0% to 14%), and an increase in old lodgepole pine (0% to 4%). This trend may be attributed to fire suppression policies and may have significant implications for increased susceptibility to insects and fire. Although it is not uncommon to find Douglas-fir in the old seral stage (>140 years), due to the natural rotation age of the species, it is not common to find lodgepole pine or interior spruce in the old seral stage in this area of British Columbia.

Results indicate that interior spruce, lodgepole pine, black spruce, subalpine fir and deciduous species were not represented in the old seral stage in the 1947 data. This trend likely reflects of the natural rotation ages of the species. During natural successional processes these species are not generally represented in old seral stages, whereas Douglas-fir is known to be a component of the older forest. This trend may also be an artifact of different classifications used on the 1947 and 1999 data. When classifying aerial photography it is difficult to discern the difference between mature and old seral stages, especially in the older, poorer quality photography from 1947.

Landscape Configuration – In addition to observed changes in landscape composition, changes in landscape configuration were observed when vegetation cover type and seral stage were combined. Most notably, harvesting focused on mature Douglas-fir, interior spruce and lodgepole pine forests has resulted in a substantial increase in fragmentation in remnant patches. In Douglas-fir, interior spruce and lodgepole pine leading stands patch density has increased in the mature seral stages, indicating an increase in fragmentation.

In Douglas-fir leading stands patch density has increased in mid seral stages and has increased considerably in old seral stages, indicating an increase in fragmentation. In Douglas-fir leading stands, patch density has decreased in the early seral stages. In interior spruce leading stands, patch density has increased in all seral stages, reflecting an increase in fragmentation in all seral stages. In lodgepole pine leading stands, patch density has increased in late seral stages and decreased in early and mid seral stages.

Changes in mean patch size (ha) were variable. In Douglas-fir, interior spruce and lodgepole pine leading stands mean patch size has decreased in mid and mature seral stages, indicating an increase in fragmentation in these seral stages.

In Douglas-fir leading stands, mean patch size has decreased in all seral stages, indicating an increase in fragmentation in all strata. In interior spruce leading stands, mean patch size has increased in early and old seral stages indicating a decrease in fragmentation in these seral stages. In lodgepole pine leading stands, mean patch size has increased in old seral stages indicating a decrease in fragmentation in these seral stages.

Changes in mean shape index were also variable. In Douglas-fir leading stands, mean shape index has decreased in all seral stages, indicating a decrease in edge. In interior spruce leading stands, mean shape index has decreased in mid and mature seral stages and increased in early and old seral stages. In lodgepole pine leading stands, mean shape index has decreased in early, mid and mature seral stages; and increased in old seral stages. Co-variation of ENN increased in early and mature seral stages reflecting an increase in patch isolation. Co-variation of ENN decreased in mid and old seral stages. In interior spruce leading stands, co-variation of ENN increased in early, mid and old seral stages and decreased in mature seral stages. In lodgepole pine leading stands, co-variation of ENN increased in all seral stages.

Changes in Road Length

Road length on the JPRF has increased by approximately 95%, from 9.1 km in 1947 to 179.4 km in 1999. In 1947 roads were located in the southern portion of the JPRF associated with the Pinchi Lake Mercury Mine. In the years following, road expansion occurred throughout the JPRF.

CHANGES IN LANDSCAPE COMPOSITION AND CONFIGURATION

Over the past 60 years on the JPRF, there have been changes in landscape composition and configuration. These changes are thought to have occurred primarily as a result of industrial management in the form of timber harvesting and road construction. There has been a decline in mature seral stages in Douglas-fir, interior spruce and lodgepole pine leading stands likely resulting from timber harvesting in these forest types. The increase in old seral stages may represent unproductive or inaccessible stands that were not harvested, and the increase in young seral stages may be the result of increased regeneration (natural and artificial) in young forests. Over the past half century, seral stage distribution is moving towards an increase in the youngest and oldest age classes. These, and other changes, changes may have significant implications for future ecological function including changes in wildlife habitat, risks of insects and disease, and changes in fire risk.

In areas on the JPRF where diameter-limit harvesting occurred, stands were frequently converted to subalpine fir as a result of the release of this shade tolerant species by partial cutting. These types often regenerated into mixed stands, especially with a deciduous species as a component. Following the 1970s, some of the stands that were harvested earlier using diameter-limit harvesting were later clearcut and frequently were artificially regenerated with interior spruce, and to a lesser extent lodgepole pine. The decline in lodgepole pine in the old seral stage may be partially the result of a successional shift in mixed stands from lodgepole pine leading stands to the longer-lived Douglas-fir as the leading type. Also, as lodgepole pine was primarily located in the north and eastern portion of the forest, these types may have been cut down along with Douglas-fir which was a highly valuable species at that time. This would be especially relevant as utilization standards became more inclusive, and therefore species such as lodgepole pine became increasingly valuable and were therefore harvested.

While the increase in young interior spruce indicates the natural and artificial systems employed to regenerate this species were successful, the decline in young Douglas-fir may reflect the difficulty of successfully regenerating Douglas-fir at the northern extent of its range. This decline may reflect fire control in the area and the difficulty regenerating Douglas-fir in the absence of fire.

Since 1987, licensees in British Columbia have legally assumed the responsibility for the establishment, maintenance and production of a free growing stand. In this context, a free growing stand is a plantation that is not impeded by brush, is healthy with respect to insects and disease, contains advanced regeneration with the potential of becoming sound and merchantable timber at the time of rotation, and meets a minimum height requirement (British Columbia Ministry of Forests 2000). The decline in young Douglas-fir stands on the JPRF may reflect policies for free growing stands implemented by the British Columbia Ministry of Forests. In this region of British Columbia, it has been common to reforest with lodgepole pine or interior spruce, as opposed to Douglas-fir, as these species (especially lodgepole pine) have faster initial growth rates than Douglas-fir and generally perform better when planted, therefore reach a free growing stand faster (British Columbia Ministry of Forests 2002). The reluctance to reforest with Douglas-fir also results because of the limited technology to process Douglas-fir in this region of British Columbia. This move away from Douglas-fir as a merchantable species is also related to reducing management risk of the species that is prone to attack by the *Dendrotonus* bark beetle.

On the JPRF, fragmentation has increased in the coniferous forest over the last half century. As well, the coniferous forest is composed of more simplified patches with less edge which have become increasingly isolated. An important result of increased fragmentation in mature coniferous forests is the conversion of interior forest to smaller patches with an increased amount of edge. Industrial land use, and the creation of smaller patches, creates high contrast edges which may influence adjacent stands for tens to hundreds of meters. Increased edges may obstruct species dispersal, divide the population and increase competition in the patches that remain. In Douglas-fir forests, smaller patches experience increased rates of windthrow, reduced humidity and altered hydrological patterns. An increased amount of edge may also impact tree growth rates, elevate rates of tree mortality, reduce stocking density and impact conifer regeneration. Although this study shows an increase in fragmentation on the JPRF (which in turn results in an increase in edge effects), the mean shape index decreased, indicating a decrease in patch complexity and fewer edges. In essence, this means that although there has been an increase in the number of patches, each individual patch has become less complex, and therefore the total amount of edge has decreased. This result may be, in part, a function of straight edges on clearcuts.

The quantity of mature coniferous forest habitat on the JPRF has declined and the remaining habitat has become increasingly fragmented. One species in particular, the American marten, is known to inhabit the forests of this region and is negatively impacted by the changes that result from the conversion from mature, contiguous forest to younger, more fragmented forest. This species selects mature, contiguous forest habitat for predator avoidance, special habitat features and the abundance of prey. The changes in this habitat type that have occurred over the past 60 years on the JPRF may have negative and lasting impacts for the American marten, and other interior forest species.

In some forests road development has been shown to leave a more lasting impact on the landscape condition than changes that have resulted from timber harvesting. In this study it was shown that between 1947 and 1999 there has been a 95% increase in roads constructed to facilitate timber harvesting on the JPRF. Although these roads are still obvious on the forest, some of the roads have become overgrown with vegetation and are impassable to vehicular traffic. Nevertheless, an increase in roads potentially leads to increased fragmentation, increased natural geological process (*i.e.* soil erosion and sedimentation), increased access for recreational activities such as hunting, and changes in species dispersal patterns. Any ecological restoration initiatives on the JPRF must consider the impact of roads on the ecological integrity of the forest. Because of the nature of the operations on the JPRF, however, these considerations must be balanced with the need for access onto the forest for social, economic and educational purposes.

Limitations of the Data

The scale and variable quality of older aerial photography impacts the quality of interpretation possible when using this type of data, and the quality of the available photography is the primary constraint on applications using historical aerial photographs to interpret past landscape structure. Similar limitations exist with current inventory data. Integrating aerial photography data with other forms of historical information helps to address this limitation. To test the quality of the database created from the 1947 aerial photography, polygons were selected and ground sampled. Since 1947 over 40% of the JPRF has been harvested, and many remaining areas that were not harvested have limited access and were not sampled. Therefore, only 16 polygons were sampled throughout the JPRF with three plots each, to represent the diversity of types identified in the 1947 database. Due to the limited number of samples the sampling was not expected to provide statistical accuracy but rather to give an indication of database quality.

From the 1947 data, leading species was correctly identified in 62.5% of the polygons. In those polygons where leading species was incorrectly identified differences may be explained by successional changes over the past 60 years, or differences in ordering of leading and secondary species. Age classes were correctly identified in 50% of the polygons. The incorrect polygons underestimated by one age class. Height classes were 75% accurate. In the 1999 database 48% of polygons were accurately identified with respect to species composition. Similar to the 1947 data, in polygons incorrectly identified, differences may be the result of successional shifts, or incorrect ordering of leading and secondary species. Only 25% of the age classes in 1999 database were correct, with the remainder showing inaccuracies of 22 years. It is important to consider these inaccuracies when making assumptions about changes in landscape structure.

NATURAL RANGE OF VARIABILITY

The primary limitation of this work involves the use of a single point in time (*i.e.* the year 1947) to reference the landscape condition on the JPRF prior to the initiation of industrial management. As the landscape has undoubtedly changed over time in response to natural disturbances, using only a solitary point in time to reflect this level of variation is a limited approach. Prior to the initiation of commercial harvesting and road construction in this area, the major natural disturbance regime was wildfire, and to a lesser extent the occurrence of insects and disease. Using multiple points over time to quantify the range of variation that naturally occurs in forests, and comparing the current condition to this natural range, would assist in interpreting the changes reported here. Otherwise, it is not possible to determine if the changes that occurred as a result of industrial management are within the natural range of variation for the JPRF, or if the changes are outside this natural range. However, as we only had access to one set of historical aerial photographs, developing a range of pre-industrial management conditions was not feasible using this approach. Many studies of this nature are plagued by these limitations.

In general, there is a lack of established values for determining the natural range of variation in ecosystems, and therefore a lack of a framework with which to interpret changes in landscape structure. To partially overcome this limitation in this study, the 1947 and the 1999 data was compared to recommendations for natural range of variation from the Biodiversity Guidebook (British Columbia Ministry of Forests and British Columbia Environment 1995). The Biodiversity Guidebook presents an ecosystem based approach to the conservation of biodiversity in managed forests. The Biodiversity Guidebook attempts to conserve a variety of natural habitats at the landscape level and presents forest management practices at the stand level that sustains key ecosystem processes. A basic tenant of the Biodiversity Guidebook is the more a managed forest resembles the historical forest which existed prior to European contact, themore likely natural species and processes will be maintained. As such, the Biodiversity Guidebook offers values for the range of natural variability for some indicators of landscape structure, namely seral stage distribution and the arrangement of patch sizes.

The JPRF is located within the SBS biogeoclimatic zone which is characterized by ecosystems with frequent stand-initiating events, or Natural Disturbance Type 3. Where conifers dominate the SBS, the mean fire interval was ca. 125 years, with remnant Douglas-fir stands remaining following these natural disturbances. The impact of insects and disease in this ecosystem type was highly variable. Based upon an understanding of these natural disturbances patterns, the Biodiversity Guidebook provides a range of natural variation for seral stage distribution and the distribution of patch sizes.

When the distribution of patch sizes for 1947 and 1999 vegetation types were compared to the range of patch sizes from the Biodiversity Guidebook, it was determined that neither the historical or contemporary data bases fell within the recommended range of variation from the Biodiversity Guidebook. Similarly, when the distribution of patch sizes for seral stages, and the combination of seral stages and vegetation cover types, are compared to recommendations from the Biodiversity Guidebook the historical and contemporary data do not fall with in the range characterized by the Biodiversity Guidebook. For instance, both data sets have patches that >250 ha in size, whereas the Biodiversity Guidebook does not include recommendations for these larger patches. This pattern would indicate that the historical conditions from 1947, nor the contemporary conditions from the 1999 forest are not within the (generalized) natural range for this ecosystem. The recommendations provided by the Biodiversity Guidebook are only a guide, and it is reasonable to predict that the data from 1947 would fall within the natural range of variation for that forest. This issue emphasizes the need to obtain site-specific data to define natural, or at least historical conditions, for the study area and not to rely solely on recommendations provided by the Biodiversity Guidebook.

The seral stage distribution for both the 1947 and the 1999 data fall within the recommendations from the Biodiversity Guidebook (1995) for maintaining high levels of biodiversity. This indicates that with respect to seral stage distribution, changes that have occurred on the JPRF between 1947 and 1999 resulting from timber harvesting and road construction have remained within the natural range of variation for this forest.

Species composition is one component of natural landscape change that has not been quantified within the context of the Biodiversity Guidebook. Without this frame of reference it is difficult to determine whether the changes in species composition that have resulted over the past 60 years on the JPRF are within the natural range of variation for the ecosystem. It is recognized, however, that the changes in species composition that have occurred on areas with high timber value are of a greater magnitude than those areas that were not as accessible and therefore not harvested.

For example, in the northern portion of the forest, near Tezzeron Lake, was heavily impacted. Based on descriptions of the natural ecosystem from the Biodiversity Guidebook, data from 1947 aerial photographs, and oral and written records, these changes are believed to be outside the natural range of variation. As such, ecological restoration initiatives should be focused in areas that have undergone substantial changes, such as on the northern portion of the JPRF.

ISSUE OF TEMPORAL AND SPATIAL SCALE

The perception of natural variability in ecosystems is dependant both on the method of observation, and on the scale that the observation was made. In an ecological system, the patterns detected are a function of the scale, both in terms of the grain and the extent of the analysis. When observing landscape change the spatial scale is defined by the size of the smallest patch (grain) and by the geographic extent of the landscape, while the temporal scale is constrained by the time when the original sample (grain) was taken and by the entire time period (extent) that is being considered. These constraints limit the interpretations of the results in studies such as this as ecological patterns cannot be detected below the grain or beyond the extent that was studied.

In terms of temporal scale, the period of time during which the landscape was studied has an impact on the changes detected and the subsequent interpretations made. In this case, as the period studied encompassed the time during which the most extensive harvesting was known to have been occurring on the JPRF, and therefore the changes that were observed documenting the cumulative impacts of timber harvesting and road construction during this time were significant. Alternatively, if the landscape was studied over incremental periods between 1947 and 1999 undoubtedly the changes observed would not be as obvious. Similarly, if 10 year periods were studied over the past 200 years the conclusions would be very different. To a certain degree, the knowledge of land use practices gained from oral and written historical records offers insight into the changes that occurred over shorter time periods between 1947 and 1999. For example, from these records it is known know that intermediate utilization standards (*i.e.* selective harvesting) were used between 1940 and 1970, and clearcutting was the primary silviculture system used between 1970 and 1999. Clearly, the spatial changes that would have occurred on the landscape under these different silviculture systems would be quite different. However, without spatial data from incremental periods between 1947 and 1999 it is not possible to discuss these spatial changes quantitatively.

In terms of spatial scale, it is not possible to draw conclusions from the data beyond the smallest patch size (or polygon) studied (grain), or beyond the total geographic extent of the JPRF. As such, caution must be exercised when comparing results from this study with other published studies, as values obtained for metrics using different scales are not comparable. For example, Sachs *et al.* (1998) determined that between 1975 and 1992 forested landscapes in the interior of British Columbia were in the early stages of fragmentation. However, the landscape in question was of a different size than the JPRF, and the time periods were different, it is not possible to compare the quantitative results from Sachs *et al.* (1998) to the results reported in this work. Therefore, it is not possible to make conclusions about the changes in landscape structure in the interior of British Columbia by combining the data or comparing the metrics from these two studies. The years 1947 and 1999 were selected for analysis as data was available for the study area during these periods. It appears, however, that the late 1930s would have been a more appropriate reference period, as management activities on the JPRF affiliated with the Pinchi Mercury Mine had already begun in 1940. Thus, harvesting activities in the southern portion of the JPRF were already in progress prior to 1947. The primary area of concern for ecological restoration on the JPRF, however, is in the northern portion of the forest. As harvesting activities began in that area in the mid 1950s the results from this work will prove useful in that effort. That is, it is possible to use changes reported here to justify further work into the potential need for ecological restoration in this portion of the JPRF. Further, a range of data that reflects pre-management conditions is necessary to definitively establish trends that have industrial management on the JPRF, and to distinguish these changes from those that would result naturally.

Previous studies have shown that changes in landscape condition in forests are more apparent at larger scales, and this issue of scale is relevant when considering a relatively large area such as the JPRF. Additionally, greater and potentially more ecologically significant changes in landscape structure are evident in smaller areas composed of merchantable timber. That is, when studying changes over large areas it is possible that the changes observed become masked to a certain degree, and these are more apparent when looking at smaller areas, which are currently, or were at some time, composed of high quality timber. This trend is apparent on the JPRF. Specifically, the northern portion of the JPRF was heavily impacted by timber harvesting because of the high quality and quantity of timber, suitable terrain and close proximity to Tezzeron Lake. In particular, Timber Sale (TS) X65876, (approximate area, 2000 ha) located in the northern portion of the JPRF, was reportedly an area of high timber value. As a result timber harvesting activities on the JPRF during the 1950s and 1960s were focused to a large extent on this area. Therefore, changes in landscape structure are much more prevalent in this and other similar areas of the JPRF. An analysis at the resource management zone level provides insight into landscape changes in smaller areas of the JPRF. More specifically, performing spatial analysis on areas which have been harvested (such as on TSX65876) may result in more obvious changes in landscape structure than have been reported here, and thus highlighting the need for ecological restoration.

THE IMPORTANCE OF GRASSLANDS

The world's grasslands are dominated by grasses and forbs and have few trees. The dominant wildlife that live in this ecosystem are grazing and roaming animals. Grasslands can be either temperate or tropical. The Great Plains of North America is an example of temperate grasslands. Grasslands are different than cultivated agricultural lands. Grasslands have a great diversity of species and have evolved and developed naturally over millions of years, affected by climate. Cultivated lands focus on the growing of a few select species—such as wheat—lowering their biologic diversity.

Grasslands are also important for carbon storage. Whereas carbon is stored in trees, plants, and organic matter on the surface of the soil in forests, in grasslands it is stored in the soil. Due to human development, most of the Earth's grasslands suitable for agriculture have already been developed. The Food and Agriculture Organization (FAO) of the United Nations, as much as 70 per cent of the Earth's 3.2 billion acres (1.3 billion ha) of grasslands has become degraded, mostly because of overgrazing. Temperate grasslands are located north of the tropic of Cancer (23.5° North latitude) and south of the tropic of Capricorn (23.5° South latitude). The four major temperate grasslands (which go by different names depending on where they are located) are the plains of North America, the veldts of Africa, the steppes of Eurasia, and the pampas of South America. These areas are dominated by grasses, with a few trees, which are limited by drought, wildfire, and grazing. The principal places that trees, such as cottonwood, oak, and willow, are found are along river valleys because they need a reliable water source nearby.

Flowers are commonly found among the grasses. The most common grass types are buffalo grass, purple needle grass, blue grama, and galleta. The soil in temperate grasslands is rich in nutrients because of the growth and decay of deep, dense grass roots. The roots serve to hold the soil together as a cohesive unit. In fact, some of the world's most fertile soils are found in the grasslands of the eastern prairies in the United States, the steppes of Russia and Ukraine, and the pampas of South America. Tropical grasslands are located near the Earth's equator, between the tropics of Cancer and Capricorn.

These are the grasslands of Africa, Australia, India, and South America. Tall grasses dominate these landscapes— some as tall as 3–6 feet (1–2 m). These are found in tropical wet and dry climates, hot all year long, their temperatures never falling below 64°F (18°C). Usually dry, there is a season of heavy rain, bringing annual rainfall to 20–50 inches (51–127 cm) per year. One of the biggest threats to the tropical grasslands from global warming is desertification. The Intergovernmental Panel on Climate Change (IPCC), the structure and function of the world's grasslands make them one of the most vulnerable to global climate change of any of the terrestrial ecosystems. Grasslands are naturally vulnerable to encroachment by nonnative, invasive species. As temperatures rise and precipitation decreases due to global warming, grasslands will become stressed under droughtlike conditions, causing them to become more vulnerable to wildfires. Once affected by wildfire, grasslands face additional ecosystem and habitat destruction. Following wildfires, native vegetation often does not grow back because it must compete with more aggressive, robust invasive species. In the Southwest, for example, it is a common problem for cheatgrass to take over during the regrowth phase that occurs immediately after a wildfire. In addition, global warming will also cause a gradual decline in soil carbon and nitrogen content, having a negative affect on the health of the returning vegetation. Scientists have experimented to determine how much of a carbon sink grasslands have the potential to be. Dr. Shuijin Hu, an assistant professor of plant pathology at North Carolina State University was involved in a study that showed grasslands are able to act as carbon sinks when carbon dioxide (CO_2) concentrations in the atmosphere are increased. In his study, he discovered that grasslands actually sequester carbon in the soil. Soil microbes respond to the changes in carbon and nitrogen and actually store excess carbon as atmospheric levels increase. Sequestration occurs when the amount of CO_2 absorbed by vegetation is greater than the amount of gas released by decomposing plant material.

In the study, Dr. Hu, along with Dr. F. Stuart Chapin III of the University of Alaska, Dr. Christopher B. Field of the Carnegie Institution of Washington, D.C., and Dr. Harold A. Mooney of Stanford University, focused for five years on a study site at Stanford University's Jasper Ridge Biological Preserve in California. From 1992 to 1997, two study plots were maintained—one at today's normal CO_2 levels (360 ppm) and the other at double that amount (720 ppm). Hu explained, "Other scientists have proposed that grasslands can act as carbon sinks when atmospheric carbon dioxide is elevated." The research they conducted showed that grassland soils can sequester carbon and also found a trend towards increased soil carbon under elevated CO_2 conditions. At the end of the experiment, they found that the soil core sample from the plot that had received double the normal CO_2 had stored more carbon.

Their results presented an interesting relationship in grasslands. When CO_2 is increased in the atmosphere, grassland plants grow more rapidly, while drawing nitrogen from the soil. This nitrogen depletion ends up keeping it from the soil microbes, which in turn reduces their ability to decompose dead plant material (which releases CO_2 to the atmosphere). There does seem to be a tipping point, however, where only so much carbon can build up. Once it reaches that level, the lack of nitrogen keeps additional plants from growing. Dr. Hu concludes that "Forests may be of greater potential as a long-term carbon sink than annual grasslands because trees can sequester carbon in above-ground biomass." Grasslands do, however, play a role in carbon storage.

GRASSLANDS AROUND THE WORLD

A news release issued by the French National Institute for Agricultural Research (INRA), the European GreenGrass project, a study conducted in Europe by Jean-François Soussana, determined that the least productive sites—those closest to mountains or at higher latitudes— stored less carbon than the grasses in the more productive areas located in the plains. The more productive sites, however, were grazed more and emitted more methane and nitrous oxides, making them more vulnerable to the effects of global warming. They determined that the grasslands they studied served as a moderate carbon sink. The extreme heat wave Europe experienced during the summer of 2003 gave scientists a rare opportunity to study its effects and take measurements. That summer, temperatures climbed 10°F (6°C) higher than normal and rainfall was reduced by one-half of the average in Europe since 1850. That summer's temperatures simulated temperature regimes that could be expected with long-term global warming.

A study was led by Philippe Ciais of the Laboratory of Environmental and Climate Sciences and French National Centre for Scientific Research to quantify carbon storage and gas emission on grasslands. The study determined that throughout Europe the 2003 heat wave reduced the net storage of European ecosystems by 500 million tons (454 million metric tons) of carbon, or the equivalent of three years of greenhouse gas emissions by a country the size of France. The study showed that global warming could self-accelerate. By lowering the role of ecosystems as carbon sinks, the greenhouse effect would be accelerated and lead to increased warming. Ciais also points out that there are still many uncertainties that need to be answered. Droughtlike conditions are not new to North America and the grasslands there.

After the 1930s dust bowl era, where much of the Great Plains region suffered through extreme drought conditions, other areas have sporadically been affected as well, such as the southwestern United States over the past five years. In the summer of 2000, severe drought also affected southeastern states (Georgia and Alabama), and large areas of Texas went more than 67 consecutive days without any rainfall. NASA, with the acceleration of global warming and rising temperatures, regions are likely to see more extreme dry events. Severe drought kills crops and dries topsoil, allowing it to be carried away by the wind, and evaporates lakes and other sources of freshwater, leaving everything parched. Current research is focusing on the ability to predict the onset of drought weeks, months, or even a year before it occurs so that adaptations to a warmer environment can be made. Rangelands cover about 50 per cent of the world's land area, and they are ecologically as important as the rain forests. They are also in as much danger of degradation and destruction.

The world's rangelands may play a significant role in buffering the effects of global warming. These areas are used for the grazing of cattle and sheep and are seen as a carbon sink by some global warming scientists. An international team of researchers led by the University of California, Davis, has focused its attention on the central Asian rangelands in an effort to determine whether this is true. Montague Demment, who is the director of the University of California, Davis, Small Ruminant/ Global Livestock Collaborative Research Support Programme, says, "Properly managed rangelands could have a relevance for global climate stability comparable to conserving the tropical rain forests. We may very well find that conservation and restoration of the rangelands, in the United States and internationally, have a significant impact on the world's 'carbon budget' and are critical factors in slowing global warming." The research team is looking at Asia's 647 million acres (262 million ha) of rangeland to determine how to restore them to healthy, productive, functioning grasslands. If this goal is met, they believe that they can store enough carbon to equal a 30 per cent reduction in the carbon emissions caused by all the people in Russia. They also make dire warnings that if grassland restoration does not happen, if rangeland degradation through cultivation or development takes its place, substantial amounts of CO_2 could be released into the atmosphere.

Rangelands offer vital functions for ecosystem health, including many that are listed below:

- Protects fragile soils
- Stores large amounts of CO_2
- Provides habitat for wild flora and fauna
- Acts as watersheds for large river systems
- Provides recreational opportunities
- Provides reserves for wildlife
- Maintains atmospheric quality
- Purifies air and water
- Retains nutrients for plants
- Provides medicinal plants
- Provides timber
- Provides germplasm for new and wild relatives of existing crop and pasture plants
- Provides food for livestock
- Provides a source for biofuels
- Provides fibres for cloth (cotton, wool, leather, hemp, flax, cashmere)
- Provides industrial products (waxes, dyes, vegetable fats and oils)
- Provides cooking oils (plant and animal fats and oils)

Although the health and destruction of grasslands and rangelands do not appear in the news as often as the destruction of the world's rain forests, they are in danger from the effects of global warming, as well as from human interference. Not only do they face degradation, but also desertification. The world's grasslands are also predicted to migrate northward, shifting the rich farmlands found in the central portion of the United States northward into Canada. Crop production may become more risky and difficult. As farming regions shift, there is no guarantee that soils would be conducive to the large-scale growth of agriculture needed to feed the world. By 2050, the world's climate could be so unstable that world food production would be reduced by 50 per cent.

6

Trees in Life Support Systems and Traditional Agroforestry

The image of the desert as little but vegetation-less, unbroken, stretches of sand and inaccessible terrain conditions is shattered upon entering the arid zones. There are sparsely distributed trees, beneath which are cultivated arable crops (especially during the kharif season, as agriculture is predominantly rain fed) and/or grasses, and other herbaceous flora, in long stretches interspersed among widely scattered village settlements and dhanis (a unique settlement pattern of the Thar intended for the life of agriculturists' families during the active cropping period away from the village, but nearby their fields). The Indian arid zone is a vegetated desert having a very rich floral diversity of about 682 species, out of which 131 are known for their economic uses.

SYSTEM OF GROWING CROPS

Farmers in the hot Indian arid zone grow arable crops in association with tree species. In fact, such integration of arable crops with trees in the farming systems is a unique, combined, protective-productive system that works on the principles of ecology, productivity, economics, and sustainability. These systems are now generally referred to as agroforestry. Because most trees are drought resistant, they are still able to provide fuel, fodder, fruit and other products, when the crops fail, as frequent droughts are a common phenomenon. Thus, trees have a very important place in the life of people in the arid zone of India, as they are directly related to the livelihood of inhabitants, and also provide the important service of climate moderation in an inhospitable environment.

RUN-OFF FARMING

This is a system of growing crops on harvested and stored water in the farm soils by constructing an earthen dam or a bund across the gentle slope of the farmland in the valley bottom. Shallow, gravelly and rocky uplands normally used for grazing are used as catchments for efficiently harvesting the runoff water. This is locally known as the Khadin system. The major components of the Khadin system are the water collecting area (low rocky catchment), contour bands (channels), moisture storage basins (khadin farm), water impounding mechanisms (bands, spillways and sluice) and a zone of cultivator's settlements.

In this region monsoon rainfall is scanty, erratic and generally the water requirement of most of the crop plants is not fully and properly met. With the result, cropping in the Kharif (rainy season) often becomes a gamble. To offset such a risk and to have some assured crop yield in such environment, one very thoughtfully developed traditional system known as Khadin, has been followed in this region for centuries (Kolavalli and Whitaker, 1996). This system makes the use of shallow gravelly and rocky uplands as the catchments for efficient harvesting of runoff water, which otherwise normally used for grazing. The water is trapped in the low lying valley plains that are converted into farmland where either the kharif or the rabi (winter) crops are raised, depending upon the amount of rainfall and consequent runoff received during the monsoon. The system takes into account not only the vagaries of the climate to raise the crop with such a low rainfall but also prevent any soil deterioration.

AGRI-SILVICULTURE

This is the most popular system in the areas receiving rainfall between 200-400 mm/yr. People protect the naturally germinated seedlings of useful trees like P. cineraria, Z. numularia, Tecomela undulata, etc., which come up in the crop fields. Mixed cropping of pearl millet, moth bean, cluster bean and sesame is carried out under these trees. These trees do not compete with the crops but rather are complimentary in terms of improving the microenvironment. These trees give fodder, fuel, fruit etc. even when crop fails during drought, thus provide sustenance to the farmer during drought. The P.cineraria and Pearl millet agroforestry is well known around the world. Some other aspects of this system are discussed with the Non wood forest products.

SILVI-PASTURE

In the arid zone there are areas where rainfall is below 200 mm/yr and with this rainfall food production is very difficult. On the other hand there are some grass species, *e.g.* Cenchrus cilieria, Lasiurus sindicus etc., which are very well adapted to such climate and make natural rangelands. The climax tree species like Prosopis cineraria and Zizyphus numularia naturally come up in these rangelands and make a silvipastural system. Animals like cows, goats and sheep are the part of this farming system. Traditionally the areas near holy places like temples are kept reserved for silvipasture with very well established social fencing. This is a very good example of sustainable management of common property resources that still exist in several places. The silvipasture systems are strictly used according to the land's carrying capacity. Also these systems cover large areas *i.e.* 100 to 500 km^2. These systems have low productivity but are an example of an excellent synergism between the nature and the natives and that's why they have been sustained for centuries.

NEW SYSTEMS DEVELOPED

Some of the agroforestry systems have been developed after research and on farm testing are discussed below.

Agri-Horticulture. After two decades of experiments on research farms and farmer's fields, a new agri-horticulture system of jujube (Ziziphus mauritiana) inter-cropped with arid legumes *e.g.* clusterbean/mothbean/greengram has been developed for the areas receiving rainfall of more then 250 mm. This agri-horticulture system has been found to give better and earlier production, year-round work and resilience to erratic rainfall. In this system initially the jujube saplings are raised in the farmer's nursery by stored rainwater (farm pond) and then saplings are transplanted at a distance of 5 x 10 meters irrigated twice in first two weeks; thereafter the trees receive only harvested rainwater. Arid legume crops are raised successfully in the interspaces of jujube.

It is a short duration (50-75 days) crop, sown at the onset of monsoon and harvested in September/October, *i.e.* before the full growth and blooming of jujube. This offers least competition and both are complimentary as the jujube provides a favorable environment for arid legumes while legumes fix nitrogen, and leaf fall adds organic matter to the soil. Besides fruits and wood from pruning (for fuel or fencing), leaves as fodder are the additional products from jujube. Overall productivity of the system in normal and drought years shows less variation in yield due to variations in rainfall. This system can provide economic returns of Rs 28000-37000/yr as compared to the sole cropping of arid legumes (Rs. 8000-12000) with low input supply, besides non-tangible environmental improvements which give resilience to drought.

Lay farming. In this system, on one piece of land a rotation of grasses for 4-6 years, followed by food grain crops like pearl millet or legumes for 2-3 years is followed and then the land is left fallow for 2-3 years. The whole farm is divided into parts in such a way that every year all three practices, *i.e.* grass production, crop production and fallow, are available in one or another part.

In this traditional system the rotation was designed considering the poor carrying capacity of arid soils.For that purpose first grasses are grown which improve the soil's physical and biological conditions, then good production is obtained, followed by a fallow period for natural restoration of the soil. During the fallow period several weeds come which improve the soil and the life cycle of various pests is broken. This system is gradually vanishing because of several social changes which are based on the thinking of exploiting nature rather working in synergism with nature.

Wind breaks/shelterbelts. Wind erosion, a high thermal regime and hot desiccating winds are some of the serious problems, which affect the establishment, growth and yield of crops in arid areas. A mixture of trees and shrubs planted across the wind direction help in reducing the wind speed. Evaporative demand studies conducted at CAZRI with three row shelterbelts of Acacia tortilis, Cassia siamea and Prosopis juliflora as side rows and Albizzia lebbeck as the central row were found effective in reducing wind speed and wind erosion. Micro-shelterbelts of tall growing plants have also been found quite effective in providing protection with about a 20 to 40 per cent increase in yield of vegetable crops grown with irrigation.

THE CHARACTERISTICS OF AGROFORESTRY

Agroforestry is an activity that combines production on the same plot of land, from annual agricultural activities (such as crops and pasture) and from delayed long-term production by trees (for example timber and services). This is obtained either by planting trees on agricultural land or by cropping (for example after thinning) on forested land. Plots that combine arable intercrops with forestry trees are silvoarable plots, while wooded plots with pasture under the tree canopy are known as silvopastoral plots.

Fig. Wheat and walnut agroforestry

THE ADVANTAGES OF AGROFORESTRY

Agroforestry provides a different land use option, compared with traditional arable and forestry systems. It makes use of the complementarity between trees and crops, so that the available resources can be more effectively exploited. It is a practice that respects the environment and has an obvious landscape benefit. Efficient, modern versions of agroforestry have been developed, that are adapted to the constraints imposed by mechanisation. The agroforestry plot remains productive for the farmer and generates continuous revenue, which is not the case when arable land is exclusively reforested. Agroforestry allows for the diversification of farm activity and makes better use of environmental resources. Agroforestry has interesting advantages from three different perspectives.

From the Arable Perspective

Diversification of the activities of arable farmers, with the building-up of an inheritance of valuable trees, without disrupting the revenue from those plots which have been planted. Protection of intercrops and animals by the trees, which have a windbreak effect, providing shelter from the sun, from the rain, from the wind, holding the soil in place, and stimulating soil microfauna and microflora

Recovery of some of the leached or drained nutrients by the deep roots of the trees; enrichment of the soil organic matter by tree litter and by the dead roots of the trees. Possibility of combining the interest of the owner (for an inheritance of wood) and the farm (for access to cultivated land). Possible remuneration for the arable farmer for looking after the trees

An alternative to full reforestation of arable land, permitting the continuation of arable activity on land whose arable potential is therefore conserved. The tree component can be reversed, the plot stays "clean" (free from scrub) and is easy to destump when the trees are clear felled (the stumps are in lines and few in number). In silvopastoral plots, fodder units can be available at different dates compared to full cropped plots, extending the grazing calender.

From the Forestry Perspective

Acceleration of the diameter growth of the trees by wide spacing (+80% over 6 years in the majority of the experimental plantations). Reduction of the capital cost of the plantation, by reducing the number of trees planted with no commercial future. A large reduction in the maintenance costs of the plantation, due to the presence of the intercrops. Improvement in the quality of wood produced (wide regular rings, suited to the needs of industry), because the trees are not subjected to cycles of competition and thinning.

Guaranteed follow-up and tree care due to the arable intercropping activity. In particular, protection against the risk of fire in susceptible areas, with pastoralism or with intercrops like vine or winter cereals (clear bare ground in summer after stubble ploughing). Agroforestry plantations on arable land allow the development of a quality wood resource that complements, rather then competes with, the products from traditionally exploited forests. It is especially important to produce wood that can substitute for tropical sawlogs, which will soon decline in availability and quality. The areas concerned will remain small in terms of their absolute value, but the production of wood from them could become a critical input to the European wood supply network. Tree species that are little used in forestry, but are of high value, could be grown in agroforestry systems: service trees, pear trees, common sorbs, walnut trees, wild cherry trees, maple trees, tulip trees, paulownias, etc...

From the Environmental Perspective

Improvement to the development of natural resources: the total wood and arable production from an agroforestry plot is greater than the separate production obtained by an arable-forest separate cropping pattern on the same area of land. This effect results from the stimulation of complementarity between trees and crops on agroforestry plots. Thus, weeds, which are spontaneously present in young forestry plantations are replaced by harvested crops or pasture; maintenance is less costly and environmental resources are better used. Better control of cultivated areas of land: by substituting for arable plots, the agroforestry plots contributes to diminishing the cultivated area of land. The intensification of environmental resource use by agroforestry systems is not resulting in more crop products.

Creation of original landscapes that are attractive, open and favour recreational activities. Agroforestry plots have a truly innovative landscaping potential, and would improve the public image of farmers to society. This will be particularly the case in very sparsely wooded areas, where plots are developed by planting arable land, and in very heavily wooded areas, where plots are developed by thinning the existing forest.

Counteract the greenhouse effect: constitution of an effective system for carbon sequestration, by combining the maintenance of the stock of organic material in the soil (the case especially with meadows), and the superimposition of a net fixing wooded layer. Protection of soil and water, in particular in sensitive areas.

Improvement of biodiversity, especially by the abundance of "edge effects". This in particular, permits a synergistic improvement, by favoring the habitat of game. The integrated protection of crops by their association with trees, chosen to stimulate the hyperparasite (parasites of parasites) population of crops, is a promising way forwards.

These favorable characteristics are as coherent with the many objectives of the laws guiding agriculture and forestry, as they are with the directing principles of the Common Agricultural Policy.

ARID ZONE AGROFORESTRY

Drylands cover more than 60% of the earth's surface. Arid zones are described as a part of the drylands, and have the most severe edapho-climatic conditions. The major distinguishing feature for defining and for planning for the arid zone is the low rainfall (below 500mm or Aridity Index <0.20) with more than 50% inter-annual variability. This makes a great difference in terms of the nature of the ecosystem, the socio-economic environment and the challenges for sustainability. High wind and solar regimes further increase the effect of rainfall variability and the whole complex makes a fragile ecosystem in which small disturbances may cause great loss to the sustainability, which are sometimes irreversible. The hot arid zones of the world are economically and environmentally disadvantaged, with unique problems.

The ecosystems of these zones are highly fragile with large risks that cause severe impediments to development programmes. Of the total land area in the world, arid zones cover 18.8%. The hyper-arid zones account for 22.3% of the total arid zones. These arid zones are diverse in terms of climate, soils, vegetation, animals, and the lifestyles and activities of the people. Little but variability in rainfall and the presence of distinctive periods of drought are the characteristics of the arid tropics. Often, the terms drought and aridity are used incorrectly. A drought is a departure from average or normal conditions in which shortage of water adversely impacts on the functioning of ecosystems, and the resident population of people, whereas aridity refers to the average conditions of limited rainfall and water supplies, not to the departures there from. In general, arid zones are characterized by pastoralism and little farming but there are always exceptions.

Arid zones are found in almost all the continents of the world. In North America, arid zones are found in the southern US and Mexico, and in South America they are located in Brazil, Argentina, Chile, and in some other countries. In Africa, substantial parts of Saharan Africa, Ethiopia, and Namibia, are classed as arid environments. The hot arid regions of India lie between 24° and 29° N latitude, and 70° and 76° E longitude, covering an area of 31.70 million hectares, and involving seven states: Rajasthan, Gujarat, Punjab, Haryana, Andhra Pradesh, Karnataka, and Maharashtra. In total, 11.8% of the country is under a hot arid environment. The arid regions of Rajasthan, Gujarat, Punjab, and Haryana together constitute the Great Indian Desert, better known as the Thar Desert, which accounts for 89.6% of the total hot arid regions of India, and constitutes the principal hot arid zone of the country.

Table : Distribution of arid regions in different states of India.

State(s) *(10^6 hectares)*	*Area* *total*	*Percent of*
Rajasthan	19.61	61.0
Gujarat	06.22	19.6
Punjab and Haryana	02.73	09.0
Andhra Pradesh	02.15	07.0
Karnataka	00.86	03.0
Maharastra	00.13	0.4
Total	31.70	100

In arid zones, vegetation is typically sparse, and is comprised of perennial and annual grasses, other herbaceous plants, shrubs and small trees. The native plant species have adaptations that enable them to reproduce, grow and survive in the most inhospitable edapho-climatic conditions. Some plants have evolved special root systems, while others have unique leaf characteristics that allow them to withstand prolonged periods of drought. The number of tree species is very limited in arid zones, and in general, they are very slow growing due to limitations of environmental conditions, but nowhere in the world are they are so intricately associated with the life of human beings.

To evade or minimize the adverse affects of frequent droughts, the native peoples in arid zones have often developed production systems in which woody perennials have a very important role, both from a productivity as well as a resource conservation point of view. Their centuries-old experiences, under diverse socioeconomic and cultural conditions, were passed from generation to generation and people have established well-contained systems of production (silivipastoral/agrisilvicultural/agrisilvipastoral/livestock husbandry) that accrue maximum benefits from their woody components. When arid zones have been looked at from the perspective of forestry development, the focus has been on the management of trees and shrubs that are either native to a particular arid zone, or have been introduced, especially for conservation purposes.

However, concern for trees and forests has increased in arid zones as it has in many other regions in recent years. This concern has been fuelled mostly by the disappearance of large areas of woodlands and trees from such fragile ecosystems in many countries of world. In the arid zone, closed forests are seldom available. In India, traditional animal husbandry and agroforestry practices have been used to manage parklands, rangelands, reserved silvipastures near holy places, lay farming, and run-off farming (traditional watershed management). Trees are managed mainly for their non-wood forest products (NWFP) and their environmental services. Animals are an essential part of the production system. The whole system has been traditionally developed to spread the risk of drought in diversified components and efficient utilization of scarcely available natural resources.

ARID ZONE AGROFORESTRY AND FOREST MANAGEMENT

As stated above, the prevalent form of forest management in arid lands is agroforestry, which is concerned with the management of trees and shrubs for conservation and sustainable development. Agroforestry in arid lands differs in many ways from the timber forestry practiced in moist tropical and other temperate ecosystems and in many cases its applications are broader in their scope. These applications include production of wood for fuel and poles; horticultural practices for fruit and nut production; range management for fodder and forage production; forestry to modify microclimate for improved agricultural crop production; and protection forestry on land susceptible to wind and water erosion.

Many rural people combine these applications of agroforestry with other land uses linked to their needs and social values. The value of forestry in the arid zone is difficult to evaluate in monetary terms. Arid zone agroforestry therefore needs to be understood broadly as management of trees and shrubs to improve the livelihood and quality of life for rural people in the arid environment. It is in this diverse context that policy developemnt must take place if sustainable development of forest resources in the arid zone is to be achieved.

Agroforestry is being followed traditionally in the arid zone to minimize the risk of drought and sustain life in climatic adversities. Several need-based systems have developed traditionally and some have been improved or incorporated in recent times.

FUTURE DIRECTIONS FOR ARID ZONE AGROFORESTRY

Every ecosystem has some buffering capacity that enables it to remain unchanged for some time if it is overexploited.

While this applies to the arid zone, its fragile nature limits its buffering capacity. Thus, for a few decades the shortcomings of policy were not observed but as the buffering capacity is gradually being eroded, the indicators of unsustainability as a whole, *e.g.* decreasing natural resources, deforestation and desertification, become visible. For fragile ecosystems, like those that occur in arid zones, one thing needs to be recognized by policy makers and scientists, that is; sustainability of agroforestry in the arid zone depends on the degree of synergism with '3Ns' *i.e.* Nature, Native vegetation and Native people.

Some of the future directions for arid zone agroforestry need to focus on the evaluation, improvement and encouragement of sustainable traditional agroforestry systems, exploring the possibilities of natural regeneration/ rehabilitation of arid zone ecosystems, NWFPs as the basis of sustainable livelihood, rethinking of research priorities, and policy improvement. Evaluation, improvement and encouragement of sustainable traditional agroforestry systems that have been developed for centuries in harmony with nature. Agroforestry systems have been developed and used in this region since time immemorial. This may be because the inhabitants of this region found that no other option was sustainable aside from conserving and harvesting whatever nature is giving. The present forms of traditional agroforestry developed over the centuries through the experiments of earlier generations in all types of climatic variations. The validation of their sustainability is very difficult to establish by conducting short-term experiments lasting only 5-6 years. Thus, there is immense possibility for evaluation, improvement and encouragement of these traditional agroforestry systems.

Exploiting the possibilities of natural regeneration/rehabilitation of arid zone/arid zone agroforestry as an alternative to failed plantation/rehabilitation strategies. It is a mechanism of restoration or maintenance of ecosystems by nature that trees produce numerous seeds. A small part of this seed population is sufficient to replace the loss of plant population due to various factors. Human beings propagate the plants artificially to compensate for what they harvested form the natural vegetation or to get a higher quantity of the natural products. Those needed in large quantities have already been domesticated, *e.g.* food crops, fruit trees etc. Still a large number of natural species are wild or depend on natural regeneration. Trees in agroforestry may fall under this category. Natural regeneration as well as artificial tree establishing technologies works comparatively better in humid and irrigated areas. In the arid zone, due to climatological, physiological and sociological reasons artificial establishment has not achieved the required success.

The need for domestication and technology development for sustainable harvesting and value addition for non-wood forest products (NWFPs). Unlike the humid or sub-humid regions where trees are typically grown for timber, the trees are grown in this arid region mainly for non-wood forest products (NWFPs). In the arid zone closed forests are seldom available and agroforestry has been followed since time immemorial in which people conserve and manage whatever seeds of native tree species like Prosopis cineraria and Zizyphus numularia germinate in the crop field or on the common land. After the establishment of trees only annual lopping of branches is done which provides precious fodder during scarcity/drought. Taking the example of P. cineraria, this tree is found in this region at densities ranging from 40 to 120 trees/ha.

This we can say is the natural forest in common or private lands otherwise no dense forest is found as in the high rainfall areas. People lop the branches of this tree in the month of December and it becomes lush green in the month of May-June when no other fodder is available. Simultaneously, pods develop which are used as vegetables. Similarly other tree species provide fodder, pods, fruits etc. The interesting thing is that, during drought when annual crops fail, the yield of NWFPs does not decrease drastically, thus trees provide food security during drought. Besides some trees produce other products like gum, resins, edible/non-edible oils etc. which increase the income of the marginal farmers who are in the majority in this region.

The services provided by the trees in hot arid regions are equally important for moderating the climate. Cultivated in the form of shelterbelts, they reduce wind erosion and conserve soil and water.

In agroforestry systems, trees reduce the solar heat load on crops and also decrease wind speed and soil temperature, conserve soil moisture, increase biological activities in the soil and thereby improve the microenvironment for crops.

The yield of crops grown under P. cineraria is generally higher than sole crops, knowing this fact agroforestry has been followed in this region for centuries. The edapho-climatic conditions in this region are so harsh that every tree improves the environment, it may be at micro level, yet its impact is visible.

The other important NWFPs in these arid areas are plants with medicinal values. Due to climatic specialty there are medicinal plants which are either endemic to this region or their quality deteriorate if they are grow/found in higher rainfall areas.

Research Priorities: Considering the social and ecological demands, policy on agroforestry research also needs to be prioritized. Some of the possible priorities are-In-situ rainwater conservation for tree establishment and for agroforestry more generally.

Water is the most limiting factor for productivity and sustainability in the arid zone. As described earlier, the effect of bringing additional water through canals or ground water has been the cause of unsustainability in the ecosystem. Thus the only possibility remains the conservation of rainwater. Trees are the keystone of arid zone conservation and once trees are established the other conservation processes start automatically *e.g.* rainwater conservation, development of vegetation cover etc.. Further research is required on:

(i) Development of in-situ rainwater conservation techniques that require less labour and maintenance;

(ii) Development of plant-based water absorbing/retaining materials;

(iii) Initiatives for widespread adoption of rainwater conservation techniques;

(iv) Improvement in traditional water harvesting systems;

(v) Development of simplified windbreak establishment techniques.

Improvement of Policy: There are some important aspects on which improvement of policy is needed for creating a more congenial environment for development. These include:

Increased international funding for arid-zone research and development. The arid regions of the world in general suffer from a vicious cycle of low productivity, a low level of investment and as result, poverty. Greater priority is given at the international level for funding of projects and programmes related to timber forestry and rainforest management, perhaps because these activities contribute more to international trade. Very limited funds aimed at combating desertification are available to the arid lands.

Forestry and agroforestry activities in arid regions no doubt contribute less to international trade but they provide substantially support to the livelihood of people in these climatically less favored regions.

There is a need to rationalize these priorities and for the international community to give due importance to the non-tradable benefits of forestry/agroforestry, which are otherwise the basis of sustainability and livelihood. Suggested areas for increased investment in the arid zone are as follows:

(i) *Organic farming in virgin arid lands*. The philosophy of the traditional systems was to work in synergism with nature and use locally available natural resources. High input agriculture (HIA) on the other hand is only sustainable with assured input supply, *e.g.* irrigation, fertilizers etc, which is not always possible in drylands. Thus the chances of failure are much greater than in traditional systems during drought. Also due to less use of synthetic chemicals in the past, these lands are comparatively free from the residues of such chemicals. Hence organic farming, based on traditional systems and conservation of local natural resources can be one area for investment without much effort. Organic farming of NWFPs can be the first choice for investment.

(ii) *Eco-tourism*. Investment in eco-tourism can prove to be a large income generator in the arid lands. Increasing numbers of people are spending increasing amounts of time and money enjoying the unique resources of these arid land environments, *e.g.* spectacular scenic beauty, diverse wildlife population, relatively clean air and other amenities such as large open spaces. Investment in the forestry sector can be in amenity plantations, desert parks, buffer strips, roadside plantations and green belts can be economically as well as environmentally beneficial for the arid zone.

Exploring the potential of arid zone agroforestry as a tool for solving environmental problems.

Conservation of Biodiversity: Biodiversity is the essential safety mechanism of any ecosystem and as the arid zone is considered to be fragile in nature the importance of conservation of biodiversity is further increased. Because of biotic interference and developmental activities, not only is the population of various species reduced but also their natural habitats have been destroyed. Thus the easiest way to save these species is ex-situ conservation with protection. This is possible through arid zone agroforestry, in which all types of plants *e.g.* trees, shrubs, climbers etc can be accommodated, combining conservation with domestication for increased production.

Desertification Control: Desertification is the result of exploitation of arid zone ecosystems beyond their carrying capacity. The carrying capacity of the arid lands is inherently low thus they are highly prone to desertification. Through soil and water conservation and sustainable management of vegetation, the carrying capacity of natural resources can be enhanced.

Various arid zone agroforestry models, *e.g.* wind breaks, sand dune stabilization, silvi-pasture, agri-horticulture etc. can be appropriate for soil and water conservation in arid lands, and the productivity of lands suffering form desertification can be enhanced and in this way desertification can be controlled. Arid lands can be considered simulated desertified drylands for planning of desertification control.

This can be important for preventive strategies for drylands rather than to control desertification after the process starts. Any technology or strategy that is found suitable to increase the productivity of arid lands, would give better results if applied to the desertification-prone drylands because they have a better environment than the arid lands. Assessing the potential of arid zones as carbon sinks.

In addition to their other benefits, trees on farms or in pastures also store carbon. Arid lands cover almost one forth of the earth's surface and any effort to increase tree populations substantially contributes to their value as carbon sinks. Due to their perennial nature and high root turnover, grasses can also prove to be a good carbon sink.

Silvipasture systems are prevalent in most of the arid zone and when both the tree and grass components are properly managed, an increase in net carbon storage can be achieved.

The arid lands are prone to desertification and trees and shrubs can be grown on these lands to prevent further soil degradation and to produce biofuel. These trees also enhance soil organic carbon content and thereby increase soil carbon storage. About 104 million ha land area in the world is affected by strong and extreme soil erosion.

CHALLENGES FOR SUSTAINABLE ARID ZONE AGROFORESTRY

Agroforestry has been followed since time immemorial for sustainable livelihood in the arid zone, but in the recent past some changes have occurred, due to which the need is being felt to develop agroforestry systems, which may be able to meet the requirements of sustainable livelihood in the future.

HEAVY BIOTIC PRESSURE

Chalcolithic and Iron Age maps of the Indian subcontinent suggest that human occupation of the current Thar Desert region was thin during these earlier ages, and that populations depended upon hunting and limited pastoralism. From the analysis of available census records, the population of the Thar Desert region registered an increase of 490% from 1901 to 1991. The present population of the Thar is about 40 million.

Decennial variation in population from 1901 to 1991 in the Thar showed a growth rate of 186% (between 1901 and 1971) as compared with 132% for the whole country. The population growth rate for the decades 1971-1981 and 1981-1991 were 36.7% and 30.7%, respectively. The population density is quite high compared with the global average of 6-8 persons/km^2 for arid zones. The major factor responsible for such phenomenal growth of population in the Thar is a wide gap between birth and death rates. Moreover, social values favor large families and positive sanctions for fertility outnumber negative ones.

Cropping on Marginal Lands

Of the total area of the Thar, only 1.3% is now under forest cover. The net cultivated area increased by 54.1% from 1915 to the early 1980s. This trend in cultivated area increase has further accelerated during the past decade (the 1990s), having increased by 11.7% compared with the amount in the early 1980s. As a consequence of this, barren, cultivable and uncultivable wastelands, permanent pastures, and fallow land have declined substantially. There was a significant decline in grazing land area as marginal lands were brought under cultivation in recent years (the late 1990s). Because of low and erratic precipitation, restricted to a very short period, only one crop, the kharif (monsoon) season crop, is sown each year in most of the Thar.

The common crops are pearl millet, cluster bean, green gram, moth bean and sesame. In the arid zone of peninsular India, in addition to these crops, sorghum, maize pigeon pea and other coarse millets are also grown.

It has been observed that the agricultural holdings during late medieval times were only one fifth of those found today. By 1951, much of the potential agricultural land was brought under plow, and the trend since then in already established agricultural tracts has been intensification of farming, and the expansion of agriculture into new lands that may be considered environmentally marginal.

Mechanized field operations (use of tractors) not only increase soil erosion but also reduces the naturally regenerated population of various useful species.

Unsustainable Development Policies

With the advancement of science and technology, efforts have also been made to enhance production in the arid zone. Some of the technologies like sand dune stabilization, wind-breaks and agri-horticulture have produced good results because they are developed in synergism with nature (ecosystem) and natives (people and vegetation).

But some other technologies imported from less environmentally stressed areas have proven to be unsustainable. For example use of tractors adversely affected the natural regeneration of native trees like Prosopis cineraria, and increases wind erosion. Irrigation with waters from the Indira Gandhi canal network has caused problems of water logging and, secondary salinization. Several exotic species either become invasive, like Prosopis juliflora, or didn't prove economically productive, like Faidherbia albida, jojaba etc. The overall impact of these imported technologies, along with high biotic pressure and a fragmented approach to desertification control, can now be observed in terms of loss of biodiversity, increasing desertification, and other adverse impacts. It is now imperative to think broadly and plan holistically for the future of arid zone agroforestry, the basis of sustainable livelihood in the arid zone.

Throughout the arid zones, there is no dearth of problems, but rapidly increasing desertification (some call it land degradation) is a worldwide problem. Land degradation appears to be more of a descriptive term than desertification, as it better describes the process of wind and water erosion, soil salinization, and loss of vegetation cover. It helps in clarifying relationships among climate, topography, soil structure, land use, hydrology, and vegetation. Throughout the arid zones of the world, the major factor responsible for land degradation is soil degradation.

At the world scale, the GLASSOD database indicates that in arid zones, more than 60% of soil degradation is caused by wind erosion, with human-induced water erosion accounting for 8% of the phenomenon. At the global level, 349.6 million hectares of land in arid zones are characterized by a light to moderate degree of soil degradation, and 42.9 million hectares have strong to extreme soil degradation.

Besides wind and water erosion, over-grazing, high magnitude agricultural degradation activities, over exploitation of vegetation for domestic use, *e.g.* for fuel, and faulty land use are other factors responsible for ever increasing land degradation in arid zones.

The links between land or soil degradation and vegetation in arid zones are complex. Vegetation and soils display differing resilience to disturbance. Although in arid zones vegetation communities can be readily disturbed and degraded, recovery rates are very slow in comparison with those of other ecoclimatic conditions. Vegetation in arid zones is highty sensitive to natural climatic variability. Decreased production and an ability to lie dormant are two strategies displayed by arid zone plants for survival at times of moisture deficiency. Although an adequate database is not available, two useful relationships between soil degradation and vegetation can be determined. First, vegetation cover may help to indicate the vulnerability of a particular habitat to soil degradation, and second, vegetation cover or community type or plant species may indicate that soil degradation.

AGRO-FORESTRY; ADDING TREES TO FARMING SYSTEMS

It is relatively new to think of farming systems as including trees, or to plant trees with farming systems in mind. As recently as 1982, for example, a major study of farming systems made no mention of trees except as an "off-farm" source of fuel [1]. At the same time, most tree planting projects were assuming that farmers would grow woodlots on marginal land far from their crops. Ideas are rapidly changing. A more recent study of African farming systems gives considerable attention to trees, especially in long sections devoted to soil conservation and the improvement of soil fertility 2. Simultaneously, social foresters have begun to give urgent attention to activities such as Niger's Maggia Valley Project, which planted windbreaks to protect farm La id from wind erosion. The convergence is not yet perfect: there are still foresters who seldom think of farms, and agronomists who seldom think of trees. Nonetheless, there is increasing agreement on the need to grow trees within farming systems, or even to conceive of a wider range of plants and animal mix in order to create a sustainable, productive farming system: for example, to a maize monoculture might be added Leucaena leucocephala, bananas, sisal, groundnuts, small livestock, pigeon peas, cowpeas, mangos, yams, Cassia siamea, or Acacia albida. Properly arranged, these would serve to add nitrogen and humus to the soil, prevent erosion, fence off the garden, and provide food, fruit, poles and firewood for consumption and for sale. Since the soil would be automatically replenished with nutrients and organic material, such a system could continue indefinitely.

Conceived in this way, agro-forestry is a land use strategy that would: (a) increase the output of food and other products on a given unit of land; (b) ensure that increases in output were stable and sustainable; and (c) raise farmers' cash incomes. It is noteworthy that these are exactly the conditions necessary for greater food security in developing countries.

AGRO-FORESTRY TECHNIQUES

Many of today's agro-forestry techniques will be adaptations of traditional practices. However, non-traditional problems are raised by the decline of shifting cultivation, growing pressure on land, and rapid deforestation. For example, there is new need for fast-growing, nitrogen-fixing tree species suitable for different ecological zones. In addition, ways must be found to compress larger numbers of trees, animals, and food crops on intensively managed farming systems. Intensive research is still required in this field. Some of the most promising agro-forestry techniques are outlined below.

Scattered Farm Trees

One agro-forestry intervention is simply to increase the number of trees or shrubs scattered among crops or pastures and along farm boundaries. These trees may have value greater than any crops they displace (as can be true of fruit trees). Alternatively, the trees may actually enhance the productivity of crops and pasturage by replenishing soils and helping to reduce erosion. For example, millet and sorghum may increase their yields by 50-100% when planted directly under Acacia albida. Often, trees can be planted that will serve multiple purposes; Leucaena leucocephala, for example, can be used for poles, firewood, fodder and soil enrichment.

Scattering trees across the farm will not by itself meet all domestic needs for tree products or ensure sustainable agriculture. However, this kind of agro-forestry action is most likely to make immediate sense to farmers. Benefits are diverse and in line with felt needs. New species can be introduced cautiously and cried alongside proven ones. Risks are minimized. In a number of countries, this approach is now the principal way in which agro-forestry ideas are being introduced.

Improved Fallows

Another technique that can be introduced without making radical changes in the farming systems is the improvement of fallows.

This is required by the fact that fallow periods in many areas have been substantially shortened. As a result, only shallow-rooted grasses of little value to the soil establish themselves naturally before crops must again be planted. The usefulness of short fallows can be greatly improved by planting green manure crops such as Desmodium and Tephrosia spp., Lupinus mutabilis, Dolichos, Indogofera, etc. In addition to improving the soil, such crops may yield livestock feed or food for human consumption. It is worth noting that fallows can be improved without the use of "trees". Nonetheless, this kind of action falls directly within the scope of agro-forestry as defined here.

Alley Cropping

A much more complex agro-forestry technique is alley cropping, as developed by the International Institute for Tropical Agriculture (IITA). This involves the growing of crops between closely-spaced lines of trees or shrubs, usually legumes.

To prevent competition with crops for light and water, the trees are regularly cut back and are kept pruned during the growing season. Larger cuttings are available for firewood. Leaves and twigs are spread as mulch among the crops. This mulch replenishes the soil with humus and nutrients, as well as providing a barrier against erosion. Among the tree species found to work best in IITA's trial plots are Leucaena Leucocephala and Gliricidia sepium.

The details of alley cropping systems will vary. For example, herbicides may be applied (as IITA does), or weeding may be done by hand. Mulch may be spread before the crops are planted (as at IITA) or afterwards (as farmers have tried with mulching in northern Cameroon).

Trees may be cut at ground level (as at IITA), or just the branches may be lopped off (as farmers do on Mount Kenya, leaving a bare pole that does not shade the crops). The appropriate combination of techniques in a given area will depend en such factors as the availability of labour for mulching and weeding, the presence of pests Chat might be spread by mulching, specific crops to be grown, and the farmers' willingness to make large changes in farming practice.

Mulching

Mulch may also be used outside of alley cropping systems to protect or rehabilitate land. For example, the application of branch mulch helps restore nutrients to the soil, reduces erosion by breaking the impact of falling raindrops, and encourages water infiltration by slowing run-off and making the soil more porous.

Buffer Strips

Strips of vegetation planted along the contour lines of slops can serve as "buffers" against water erosion. Perennial grasses have long been used for this purpose in many countries. In more complicated systems, buffer strips may include grasses, fodder legumes, trees and shrubs. In such cases, the strip can yield mulch, forage, fruit and fuelwood, as well as controlling erosion.

Terracing

On gentle slopes, water erosion may be prevented through buffer strips and mulching. Where slopes are more severe, however, construction of terraces may also be necessary. This job will be made simpler if buffer strips are included in the form of dense vegetation planted along the risers and edges of each terrace. If this is done, a much smaller earth structure will be required to control run-off than would the case with conventional, "naked" terraces.

Windbreaks

Planting lines of trees to break the wind sweeping across open land can prevent erosion and slow the evaporation of moisture from the soil. Species can be chosen that will also yield fodder, poles or firewood. Individual farmers may act to protect land by planting trees on the windward boundaries of their farms. More ambitious schemes seek to control wind erosion over much larger areas.

For example, a CARE project in Niger's Maggia Valley established 250 km of windbreaks in seven years. These consisted of staggered, double rows of trees, primarily Azidirachta indica and Acacia scorpioides, planted across the valley at 100-metre intervals. Roughly 2,500 hectares of farmland were protected in this way. Initial indications are that millet production has been increased by 15% above yields on comparable, unprotected land.

In contrast to most agro-forestry techniques, large-scale windbreak schemes necessarily involve whole communities. This raises delicate issues in terms of sacrifices (whose land will be pre-empted for planting the trees?) and benefits (whose land will be best protected from the wind, and who will get such by-products as fodder and firewood?). Social analysis will obviously play an important role in this kind of project.

Live Fencing

Many of the trees and crops in an agro-forestry system are vulnerable Co being eaten or trampled by livestock. Farmers will try to keep the animals away from the plants.

However, adequate supervision of the animals may be Coo time-consuming, and wire or metal fencing too expensive. To solve this problem, farmers can plant "live fences" in the form of tightly spaced lines of species such as Prosopis juliflora. Acacia scorpioides, or Euphorbia spp.

Woodlots

Farm woodlots have often been promoted to provide firewood, either for domestic consumption or for sale. As far as domestic consumption is concerned, it will sometimes be appropriate for a family to plant such a woodlot on marginal land. However, recent estimates suggest that household firewood needs may be harvested from as few as 50-100 trees. This is most likely to be true if the trees are scattered around a farm rather than packed into a woodlot, where yields per tree are lower. Any serious agro-forestry scheme would include several times that number of trees as part of windbreaks, buffer strips, or scattered farm plantings. Under these conditions, there would be little need for a separate woodlot to produce fuel.

The growing of firewood for sale will usually be less attractive. Almost everywhere in developing countries, farm-gate prices of firewood are extremely low. Under these conditions, any attempt at firewood farming will lose money, either in absolute terms or relative to alternative uses of the same land. Where woodlots are established, these are most likely to be for production of high-value outputs such as building poles or timber, although markets for such commodities are limited.

Project Components

The exact nature of any agro-forestry project will depend on the results of the rural appraisal outlined above. However, it is safe to assume Chat most projects will involve some form of research, demonstration, on-farm trials, extension, and monitoring and evaluation.

Research

Research must be done on the impact of including trees and other agro-forestry plants within farming systems. Both exotic and local agro-forestry species should be tried. Harvesting regimes should be designed to provide the products that people actually want, as determined by rural appraisals. Since many agro-forestry species grow slowly, research on these issues is a long-term endeavour. However, it may also be a long-term endeavour to convince large numbers of people to adopt complete agro-forestry systems. If research is begun immediately, results may therefore be known at just the point when people are looking seriously for proven techniques.

Demonstration

Even while research is still underway, it will be possible to demonstrate basic agro-forestry practices whose results are relatively well understood: for example, windbreaks; scattered plantings of Acacia albida, Leucaena spp., and fruit trees; live fencing; etc. Demonstration areas should incorporate crops, trees, and harvesting regimes that are consistent with local practices and desires. Management of these areas should be flexible, with species and techniques added or subtracted as they prove more or less convincing. An important part of this work would be to demonstrate agro-forestry plots alongside fields planted in traditional ways. This would allow farmers and other visitors to see quickly the differences made by agro-forestry techniques.

On-farm Trials

The ultimate test of an agro-forestry system is its performance on farmers' fields. At the farm level, results (and perceptions of results) may be very different from findings on research or demonstration plots. Agro-forestry projects will therefore need to sponsor on-farm trials of the most promising techniques. To be useful, these trials should as much as possible be conceived and evaluated by the farmers who are to carry them out. This implies the participation of farmers who are already fairly conscious of agro-forestry relationships. Results would be used to modify research and demonstration programmes, as well as to determine the agro-forestry message to be passed on by extension services.

Extension

The first extension job is to teach the extension agents themselves about interactions among trees, land, and food. Then, guidelines would be provided on how to evaluate farming systems, so that agents could guess what interventions might be appropriate under specific local conditions. Much information for this teaching programme would come from rural appraisals. Once the extension agents understood agro-forestry concepts, they could begin to introduce promising species and ideas. This would be done gradually, starting with straightforward techniques such as boundary plantings of fruit and leguminous pole trees. More complex techniques would be channelled through the system after being proved successful during on-farm trials.

Monitoring and Evaluation

Monitoring will be extremely complex, requiring measurement of yields (of both crops and trees) in very different places: research plots, demonstration areas, farmers' fields. These data are needed to assess the actual impact of various agro-forestry techniques. Equally important is to determine farmers' responses to agro-forestry. This should be judged by the number of farmers who become involved, and by what they say about their experiences.

National and Local Institutions

Each agro-forestry project would require a multi-disciplinary implementing group. A social scientist would be needed to conduct rural appraisals, to help design on-farm trials and extension messages, and to monitor project activities in terms of farmers' reactions and socio-economic results. The research programme would require joint effort by an agronomist, a forester, and an animal husbandry expert. An agro-forestry extension officer would supervise the communications support system, in collaboration with the national extension service. Additional technical and extension staff would be based at out-stations to run agro-forestry demonstrations and on-farm trials.

If the process is functioning properly, participation of farmers is guaranteed throughout. The original choice of research priorities and extension messages builds on the situation and attitudes of farmers. Farmer's' groups are identified or created to carry out on-farm trials and participate in other activities. Monitoring gives voice to farmers as the project proceeds. At the end, farmers are asked to help evaluate what has happened. Of course, this participation will not happen automatically; vigilance must be maintained to ensure that the farmer's point of view remains at the project's heart. Ideally, all of this work would be centred within the ministry already responsible for farmers and farming systems. It might prove desirable, therefore, to create an agro-forestry "cell" within the ministry of agriculture or rural development, made up of the kinds of professionals specified above.

In many countries, there exist non-governmental organizations (NGOs) that have close and continuing relationships with distinct groups of farmers. Some of these NGOs will already have experimented with tree-planting, generally as a means of meeting family needs for wood energy. Within an agro-forestry project, these NGOs could play an important role by carrying out demonstration activities and on-farm trials. However, close links would have to be maintained with the government, which presumably would retain primary responsibility for research and extension.

7

Planning and Management of Urban Green Areas

VIGOROUS GROWTH OF URBAN AREAS

The vigorous growth of urban areas has not only increased the need for intensified urban planting but also highlighted major problems related to species selection, establishment techniques, care and maintenance, and planning. Proper species selection, tree care and maintenance will ultimately result in a greater understanding of urban tree management. Consequently, this will ensure that planted trees remain healthier and live longer, thus reducing labour and replacement costs, while still providing an overall ameliorating effect. Money saved is therefore equivalent to money earned, since it becomes available for other purposes.

In Denmark, the major part of the public green area is administered centrally by a municipal park administration. The park administration is typically located in a local council's technical department and has its own operating organization that is responsible for the practical care of green areas. However, throughout Europe there is a trend towards an ever-more privatized operation of green areas, demonstrated by increasing use of private contractors, first and foremost when establishing new installations but also in the maintenance of green areas. Nevertheless, it is characteristic of issues concerning urban forests or green areas that there is a high level of public involvement. Also, self-administration is gradually gaining a foothold, for example in day-care centres, where the management and parents' committees are increasingly able to decide for themselves how their green areas should be operated.

In addition, there is a growing tendency towards reducing costs related to tree establishment and care. Urban trees and uneconomic forest stands are particularly exposed to such cuts. Therefore, better planning and management systems for urban green areas are often currently lacking. Small individual economic cuts may not be very significant in themselves, but considered over an extended period the decrease in budgets can have detrimental consequences not apparent in the isolated individual case. One approach for solving this problem could be the introduction of overall green infrastructure planning.

GREEN INFRASTRUCTURE PLANNING

The EU Green Book ascertains that there are many threats to green areas, for example from vehicles and advertising (Commission of the European Community 1990). It also stresses the need for an element of nature in the urban environment. Even if urban nature perhaps cannot compare to 'natural' nature, it is still of great importance. The nature that exists in the landscape is the source we should use for disseminating wild plants and animals into urban areas. This is why green corridors leading from the countryside into urban areas are vital, as is the transitional zone between urban and rural areas. In many places, the green structure is not cohesive and thus cannot provide the necessary transport routes. Just a minor interruption of a green corridor can prevent it from functioning. Although binding the green network together in urban areas demands a great effort, it is absolutely essential if we are to succeed in developing urban biotopes and creating the conditions needed for a richer flora and fauna. The EU Green Book recommends that public green space should be increased as much as possible. Hence, green infrastructure planning includes a holistic assessment of the green infrastructure, current conditions and plans for development that could be achieved, for instance, by making it possible for local councils to give a green plan formal status as a part of council planning.

Examples

Germany is one of the countries at the forefront of green infrastructure planning. The German nature and environmental protection legislation also regulates parks and urban green areas. Their green plans are included on several different levels in landscape planning (which is linked to general plans) and in green structure planning (which is linked to local plans). Although these plans are not legally binding they still play a significant role. Examples from three German cities illustrate the divergent ways in which these plans originate and how they are used.

In Hannover, the overall green structure is included in two planning phases. The park administration has the option of conducting a landscape analysis before work is started, in all forms of area utilization. In this way, planning takes landscape potential into account from the very beginning. Plans for the green structure are combined with the plans for building development in the next phase. Both phases are laid before the politicians, who thus gain the opportunity of seeing how consideration is given in the final proposal to the original, overall intentions. In cases where it is impossible to avoid damaging nature, the authorities can demand compensation. Advantage of this is often taken by the city head gardener to demand improvements elsewhere, for instance a builder can be obliged to undertake the improvement of a neighbouring park. In this way, nature - the green area - becomes a commodity that can be traded, for better or worse.

Stuttgart approaches superior green planning another way. This city has a long tradition of securing the expansion and improvement of its green structure through exhibitions held every 10 years, most recently in 1993. Stuttgart also offers a classical example of the fact that it is not only the lack of green areas that can present a problem; the problem can also be related to access to these areas and their lack of mutual cohesion. The major problem in Stuttgart was the cohesion of the old royal gardens in the centre of the city. For this reason 10 bridges, which linked the parks together across various arterial roads, were built in conjunction with the Bundesgartenschau of 1977. This was a phase in the long-term 'Green U' plan, which was completed in connection with the 1993 exhibition. The Green U now makes it possible to pass through green areas, from the central royal gardens to the forests at the edge of the city.

In Munich, a plan was adopted in 1992 for 14 green strips, totalling 584 ha, which are to link the city's green areas to the surrounding landscape. The green strips are to be established over a 25-year period and will cost a total of $US430 million. The reason for this major effort is the fact that the development of the city's green areas has not increased in proportion to building development. At the same time, increased awareness of the significance of green areas to ecology and recreational activities has meant that greater attention is being paid to the importance of the cohesion of green areas. A large part of the green areas are on the edge of the city and are owned by it. Only the innermost 10-m belt of this resembles a park and consists of small tree plantations, a green and perhaps a stream. It also contains a network of footpaths and cycle paths. The remainder of these areas is rented to farmers, who operate them with only limited use of pesticides.

In Denmark, a successful example of green infrastructure planning is the 'Finger Plan' for the area outside Copenhagen. The name relates to the fact that the overall structure is shaped like a hand, with the five 'fingers' being planned for urban development. The Finger Plan, which was born as long ago as 1947, had the goal of stopping the layered growth of the city and ensuring that urban growth would thereafter be concentrated in narrow urbanized areas ('fingers') along traffic corridors. The wedges between the fingers were to be used for agricultural and recreational purposes. This idea was very clear and was beneficial to the green areas. Subsequent developments have largely adhered to the guidelines of the plan.

In England the planning of the new towns was founded, even as early as 1900, on the concept of using trees and woodland as a setting for urban development. The aims of the new towns were to relieve the urbanization pressure on large cities such as London, Birmingham and Glasgow, and to create self-contained and balanced communities for living in, for employment and for recreation. The new towns were designed and developed by development corporations. The success of the new towns is to be found in the fact that they were produced as integrated structures with good communications, high levels of amenities and higher than average physical environment, based to a large degree on social commitment and the belief that good physical environment was good for people as well as for business.

Conflicts in Management

In the policy-making, planning and development of urban forestry, some general trends and challenges can be recognized. These comprise a growing complexity, an increasing role of public participation and conflict management, and the transition towards closer-to-nature forests. In a study of urban forest planning and policy-making in 16 European cities, the main conflicts were discovered to be conflicts in relation to nature conservation, forestry vs. urban development, conflicts between different types of recreation and recreation vs. conservation management.

THREATS TO GREEN AREAS

Besides the harsh growing conditions that plants experience in the urban environment, there are other, more immediate threats to urban trees and woodlands. These are the pressures of urbanization, which compete with the green areas in demand for land; hence the high recreational pressure on the areas left for urban green spaces.

URBANIZATION PRESSURE

There is great pressure on urban space resources. Even where this is a question of public buildings, such as museums, where green area remains accessible to the public, it would in many cases detract from the overall recreational quality of the area. In Mexico City, the proportion of green areas within the city is falling by about 3.7% annually.

These are often replaced with buildings, especially in the poorest quarters of the city. Traffic installations and noise are other threats to green areas. Roads can isolate green areas from each other, which reduces their recreational value and their value as corridors for the propagation of flora and fauna. The annoyance caused by noise is more indirect. Dutch studies indicate that road noise annoys about 20% of the population, whereas about 11% is annoyed by air traffic. These figures apply to indoor annoyance. Outdoors, people are exposed to even more noise, especially since part of the urban green areas consist of 'residual areas' along traffic constructions.

Social Factors

Another important urban stress factor is vandalism. This is usually perceived to be a serious factor, although surveys shows that it is not: it is usually the result of careless play and affects only the most exposed trees. Vandalism is predominantly a social problem. Successful community landscaping and gardening in densely populated inner-city neighbourhoods have shown that one deterrent to vandalism is the development of a spirit of proprietorship in residents. Nowak *et al.* (1990) noted the highest tree mortality in areas of lower socioeconomic status. Percentage tree mortality was most strongly correlated with percentage unemployment.

Economic Cuts

In addition, economic cuts are an obtrusive threat to green areas. Park administrations responsible for a large part of city green areas have been hit hard in recent years by cuts in appropriations and personnel. As an example it is typical of the trend in Denmark over the past 20 years that economic resources have dropped by 10-20% while green areas have increased by 20-40%. Similar trends have been observed in other countries, for instance Sweden, Germany and the UK. The sector responsible for the planning, establishment and operation of green areas in Denmark (including the administration of natural and recreational areas) has an annual turnover, public and private, of $US1.35 billion and employs the equivalent of 30000 full-time employees. The maintenance costs of urban green areas amount to approximately $US16 per inhabitant. In this respect, if Denmark can be regarded as an average country in European terms, this gives a total European urban greening maintenance cost of approximately $US3.84 billion. In 1991, it was estimated that at least $US300-350 million was spent annually to enhance amenity tree resources in the UK.

ENVIRONMENTAL ASPECTS

Trees influence the urban environment in various ways, whether individual trees or the entire urban forest. Urban trees can mitigate the environmental impacts of urban development by moderating climate, improving air quality, providing habitat for wildlife and reducing noise levels. The environmental benefits associated with urban trees and woodlands are widely recognized. In large cities such as Chicago and Mexico City, the environmental aspects are integrated in the city greening programmes. At the UN Conference on the Environment and Development in Rio de Janeiro in 1992, all participating countries adopted Agenda 21, an action plan that obliges them to work towards sustainable development, an obligation that in turn devolves to the administrators of urban green areas.

LOCAL-SCALE CLIMATE

The replacement of natural surfaces with buildings and roads, as typified by cities, has altered the thermal and moisture properties of the area, which modifies the local atmosphere and generates the 'urban climate', with poorer air quality and increased air temperatures. By transpiring water and shading surfaces, trees lower local air temperatures.

In areas where the summer is hot, trees are important because of the shade they provide. Trees can even reduce energy use in buildings by lowering summertime temperatures, shading buildings during the summer and blocking winter winds and consequently reduce the emission of pollutants from power-generating facilities (McPher-son & Rowntree 1993; Nowak 1995).

Trees and shrubs can be used for manipulating air movement in order to control the impact of winds, by obstruction, guidance, deflection and filtration. The resulting effects depend on plant size, shape, foliage density and retention, as well as the placement of the plants. Combinations of trees and shrubs provide the most effective barriers, which may even be created by making use of already existing landforms. Windbreaks provide protection for considerable distances, up to 20 times the average windbreak height (Jensen 1955; Grey & Deneke 1978). A medium-porous windbreak reduces wind speed nearly as much as an impenetrable barrier. Meanwhile the shelter from this type of windbreak extends over a longer distance downwind, with much less turbulence present. However, the type of optimum windbreak depends on the situation. Hence, in some settings an impenetrable windbreak will provide the most appropriate protection, whereas in others the highly porous windbreak will be the optimal solution. Thus, the structure of windbreaks in terms of suitable provenances and clones of species, number of rows and density can be designed to provide the type of shelter needed.

Air Quality

Trees intercept particulate matter and absorb such gaseous pollutants as O_3, SO_2 and NO_2, thus removing them from the atmosphere. Trees also emit various volatile organic compounds, such as isoprene and monoterpenes, that can contribute to O_3 formation in cities. Studies conducted in Frankfurt am Main have shown that streets along which trees are planted contained 3000 polluting particles per litre of air, whereas streets that lacked trees had as many as 10 000-12 000 particles per litre of air. Protective plantations along heavily trafficked roads and around industrial areas are therefore an effective means of reducing air pollution. However, this should obviously not be taken as an excuse for neglecting to combat pollution at its source. In The Netherlands, some cities operate a tree-planting plan, the objective of which is to plant as many trees as needed in order to absorb the quantity of CO_2 emitted in providing the city with heating. To attain this goal, support is also given to reforestation programmes in eastern Europe and Africa, as well as to the extensive tree-planting carried out in the cities themselves. Mexico City's Urban Forest Programme is another example of increasing awareness of the needs of urban greening as a means of altering the overall quality of life in the city. The fundamental goals of the programme are to mitigate the effects of the severe air pollution in the metropolitan area, to establish more green areas within the city and to give the city a more welcoming appearance.

Because trees can store carbon for long periods, they are an important carbon sink. Even though plants absorb CO_2 and produce O_2, it is important not to assign excessive significance to their role in the urban environment. Harris (1992) reminds us that plants really have only a minor effect on the CO_2 and O_2 content of urban air. Photosynthesis in the oceans accounts for 70-90% of the world's total O_2 production, for which reason it is absolutely vital that they be protected against pollution. However, even a minor reduction in the O_2 content of the air will cause a large percentage increase in its CO_2 content, which would reinforce the greenhouse effect, thus leading to a rise in the global temperature. Harris (1992) also stresses that urban vegetation has an especially beneficial effect on air pollution through its ability to reduce the quantity of airborne particles. Studies of 9000 trees in Chicago have shown that the trees reduce air pollution by 12 t of CO_2 and 10.8 t of O_3 per day.

Biodiversity

Green areas play a vital role in urban biodiversity because these are the main habitats of urban plants and animals. The size and location of the urban green areas are of importance for their qualities as biotopes. Vacant urban wasteland in inner cities often possess a high diversity of species adapted to urban environmental conditions and pressures. For instance, older well-established installations attract birds and mammals whose natural habitat is the forest. The urban forest structures are also of great importance, providing wildlife corridors and stepping stones that connect the habitats in the urban green areas. Swedish examples emphasize that only if the urban infrastructure is integrated with the green infrastructure is it possible to achieve biotope protection within cities.

Since a major part of Europe's population lives in urban areas and receives its daily perception of nature therein, nature in urban areas is important for environmental awareness and an understanding of nature.

Urban areas contain more of nature than is immediately apparent. Older gardens and parks, not to mention churchyards, often have noticeably rich biodiversity. Nature created by humans is often considered to be inferior to nature that evolved without human intervention. In support of this, it is often asserted that, for example, the number of species is often greater in untouched nature.

This is mainly related to the fact that many urban green areas do not boast particularly rich biodiversity. Most of them were established with large paved areas, gravelled areas, well-mown greens and isolated individual trees. However, Owen (1992) showed that in an urban private garden with variegated flower beds and a well-proportioned mixture of cultivated and uncultivated plants it was possible to attract a large proportion of the indigenous species of butterflies (34%), moth (30%) and hover flies (36%).

The presence of birds in the immediate environment is a vital element in the context of recreation. Their presence in urban areas depends on the character of the local vegetation. Variation in the structure of plantations also forms a basis for a rich assemblage of birds. As it takes many years for new plantations to mature, it is vital to preserve as much of the existing vegetation as possible when establishing new areas.

Sustainable Urban Forests

Sustainable urban forests are defined by Clark *et al.* (1997) as 'the naturally occurring and planted trees in cities which are managed to provide the inhabitants with a continuing level of economic, social, environmental and ecological benefits today and into the future'. They will provide long-term net environmental, ecological, social and economic benefits.

To the components of a sustainable forest, McPherson (1998) includes features of adequate species and age diversity, a large percentage of healthy trees that are well adapted to local growing conditions, and a climate-appropriate treecover with native forests stands as one component of overall canopy cover.

8

Forest Service and System

FOREST SERVICE

The Forest Service is a bureau within the Department of Agriculture. It was established in 1905, to provide quality water and timber for the Nation's benefit. Over the years, the role of the Forest Service has expanded, in response to public needs, to include management for multiple uses and benefits and for the sustained yield of renewable resources, such as water, forage, wildlife, wood and recreation. Multiple use means managing resources under the best combination of uses to benefit the American people while ensuring the productivity of the land and protecting the quality of the environment.

The Forest Service manages public lands, collectively known as the National Forest System. Currently, there are 155 national forests and 20 grasslands in 44 states, Puerto Rico and Virgin Islands, which encompass 191 million acres, approximately 8.5% of the total land area of the United States.

The main activities of the Forest Service include:

- Protection and management of natural resources-on National Forest Service Lands
- Research-on all aspects of forestry, range land management and forest resource utilization
- Community assistance and cooperation-with State and local governments, forest industries, and private landowners to help protect and manage non-Federal forest and associated range and watershed lands to improve conditions in rural areas
- International assistance-in formulating policy and coordinating U.S. support for the protection and sound management of the world's forest resources.

The phrase "multiple use" is very appropriate to the mission and role of Forest Service. People are allowed a wide variety of recreational activities on national forest lands, including:

- Camping

- Hiking
- Fishing
- Hunting
- Using off-road vehicles-in designated areas

Additionally, other activities of a more commercial nature are permitted, including:

- Timbering-by selective cutting and clear cutting
- Mining
- Grazing

TIMBERING

While there are several methods used for harvesting trees, the two most widely used are:

- Clear-cutting, and
- Selective cutting

Clear-cutting involves the removal of all trees from a single area in a single cutting. After clear-cutting, the area is generally reseeded, either with a single type of tree or with a combination of trees. Clear-cutting increases the timber (and therefore monetary) yield in a given area, permits reforestation with specific types of trees, and reduces the time required for establishing a new stand of trees. Clear-cutting, however, produces large scarred areas, and in so doing, reduces the use of the area for recreation and fragments wildlife habitats. If clear-cutting reduces biodiversity, it leaves the area susceptible to invasion of destructive organisms. Finally, clear-cutting leads to soil erosion, sediment pollution of surface waters and flooding.

Selective cutting involves the removal of intermediate-aged or mature trees, cut singly or in small groups, creating only small gaps in the forest. Selective cutting is used to reduce crowding and to encourage the growth of smaller trees. It maintains a diverse stand of trees, and does not severely exacerbate soil erosion. Many industry and government foresters maintain that selective cutting is not profitable, and does not promote the growth of the most sought-after trees. However, other private and government foresters have demonstrated that satisfactory profits can be attained through selective cutting.

THE NATIONAL WILDERNESS PRESERVATION SYSTEM

Wilderness areas are those where humans are only visitors and are not allowed to remain. These are areas, each generally in excess of 5000 acres, which have retained their pristine character. In 1964, Congress passed the Wilderness Act, "...in order to assure that an increasing population, accompanied by an expanding settlement and growing mechanization, does not occupy and modify all areas within the United States and its possessions...". Additionally, it was recognized that there was a need for ecosystem studies, and a need for the protection of threatened and endangered species which called for preserving wilderness areas. Unlike the National Park and National Forest Systems, the National Wilderness Preservation System is administered by a single cabinet-level department. Rather, wilderness lands may be managed by one or any combination of the following bodies:

- Forest Service
- National Parks Service
- Fish and Wildlife Service
- Bureau of Land Management

In 1997, there were 657 designated wilderness areas, which encompass more than 104 million acres, approximately 6% of the total land area of the United States. The largest single manager of wilderness areas is the National Parks Service (41%), followed by the Forest Service (34%) and Fish and Wildlife Service (20%). More than half of the wilderness acreage is located in Alaska. Designation as a wilderness area allows for only strictly controlled recreational uses. Among the uses to which a wilderness area can be put, by humans, are:

- Hiking
- Camping

Among the many uses to which wilderness areas cannot be put include:

- Vehicular activities
- Timbering
- Mining
- Commercial activities

Further, there can be no permanent or even temporary roads, no motorized equipment, no form of mechanical transport.

The designation of wilderness areas is often a highly controversial act. Wilderness designation takes large amounts of land out of use. For that reason, timbering, mining, ranching, energy and development interests are often strongly opposed to expansion of the wilderness system, wanting to be able to put these areas to more profitable use. Furthermore, many people choose not to visit wilderness areas, because of the restrictions on what they would consider daily necessities (automobiles, electricity, piped in running water, rest rooms, etc.). For that reason, many of those people do not understand the necessity of preserving wild areas and do not support the expansion of the concept. However, many other people who may not visit wilderness areas still support the need for wilderness areas, feeling that they want to know that these areas exist, even if they, themselves, do not use them. Wilderness areas provide undisturbed habitats for wild plants and animals, promoting biodiversity.

THE NATIONAL WILD AND SCENIC RIVERS SYSTEM

In order to protect rivers and river segments with scenic, recreational, scientific, historical or cultural significance, Congress, in 1968, passed the

National Wild and Scenic Rivers Act. These river segments are to be kept free of development, straightening and damming. According to the National Wild and Scenic Rivers Act, these river segments are to be permanently administered by an approved agency or political subdivision of the state or states concerned.

The National Wild and Scenic Rivers Systems actually encompasses three types of protected river segments:

- Wild rivers-those rivers or sections of rivers that are free of impoundments and generally inaccessible except by trail, with watersheds or shorelines essentially primitive and waters unpolluted.
- Scenic rivers-those rivers or sections of rivers that are free of impoundments, with shorelines or watersheds still largely primitive and shorelines largely undeveloped, but accessible in some places by roads or railroads.
- Recreational rivers-those rivers or sections of rivers that are readily accessible by road or railroad, that may have some development along their shorelines, and that may have undergone some impoundment or even diversion in their past.

The only activities to which these river segments (and their shorelines) can be put include:

- Camping
- Swimming
- Non-motorized boating
- Sport fishing
- Sport hunting

Some new mining claims are also permitted in some areas.

In 1997, there were 150 sections of rivers protected as part of this system, corresponding to approximately 3.5 million miles or 0.3% of the total length of the country's waterways. Many environmental groups are pushing to add as many as 1500 new segments of rivers to the system. This ambitious goal, however, is being strongly opposed by many private interests and some local communities. There is also a movement among environmentalists to encourage the Congress to establish a permanent federal administrative unit to manage the system.

URBAN LAND USE

Urban areas are here defined as cities or towns with populations of 2500 people or more. The term "degree of urbanization" is used to define the proportion of a nation's population living in urban areas. Worldwide, the number of people living in urban areas has tripled since 1950. Because there is such a dense concentration of humans in these urban areas, there are very significant stresses both on the people living there and on the physical environment. Some of the important topics associated with proportionally high degrees of urbanization are:

Immigration to Urban Areas-opportunities that are driving people into urban areas in most parts of the world;

Urban Stresses-both social stresses and stresses on the environment; and,

Emigration from Urban Areas-the stresses that cause people to leave, as well as the opportunities that draw people away from, urban areas.

HISTORICAL GROWTH OF URBAN AREAS

Many of us have heard about "the threat of urban sprawl"-the thought that people are flocking to urban areas in the U.S. and worldwide at an ever-increasing rate, and that consequently these growing population centers are rapidly consuming agricultural land, parks and forests, and grasslands. To understand the reality of this issue, both now and in the near future, two points need to be considered:

POPULATION GROWTH IN URBAN AREAS

The population of urban areas is increasing very rapidly. In 1900, only 14% of the world's population lived in urban areas. In 1995 that percentage had grown to 42%, and by the year 2025 experts estimate that 60% of the population of the world will be concentrated in urban areas. The urban population growth rate in developed and in less developed countries.

The growth rate of urban population in less developed countries (LCDs) is exponential. In 1950, only about 19% of the people in less developed countries lived in urban areas. In 1995 the percentage was 34%, and the percentage is projected to be at least 57% by 2025.

The urban population growth rate in the more developed countries (MDCs), including the U.S., is similar to that of LCDs only in that the percentages are still increasing.

The growth rate in urban areas in the developed countries is significantly slower, however, than that in the less developed countries. In 1950, the percentage of people living in urban areas in MDCs was approximately 56%. In 1995, the percentage was approximately 73%, and in 2025 the percentage is projected to be 84%. In the U.S. specifically, the percentages of urban inhabitants has grown from 63 % in 1950, to 75 % in 1995 and is expected to be approximately 90% in 2025.

Worldwide growth of urban areas has led to the development of mega-cities, urban areas with 10 million or more people.

In 1995, there were 15 such localities, led by Tokyo with over 26 million people. Most mega-cities are located in less developed countries.

Only two, New York City and Los Angeles, are found in the United States. By the year 2000, the number of mega-cities is projected to grow to 26, with more than two-thirds located in less developed countries.

Areal Growth of Urban Areas

Much has been written and said about the areal growth of urban areas, often termed "urban sprawl". However, despite near-explosive growth of urban populations throughout the world, the total amount of land occupied by urban areas is still relatively small.

Even in the U.S., the world's most urbanized nation, only 3 % of the total land area is so occupied. So, what is the problem?

The problem resides more than just in the amount of land occupied by urban areas. The bigger problem is the quality of the land that tends to be occupied by urban areas. Many urban areas develop in the most agriculturally valuable land, because it is important to have a central area for farmers to tend to their agricultural business. However, as urban population increases, the area occupied by the people must also increase. Therefore, the area occupied by urban settlements grows at the expense of the farmland that many were founded to serve. This problem is just as important to the developed nations as it is to the developing nations.

In 1994, the government of Taiwan decided to convert 18% of that country's farmland to factories and housing over a period of 10 years. Not only did this action break an ancient social custom; it placed Taiwan in the lead in making the shrinking of agricultural land a matter of official policy. This policy might have made some sense if there was a surplus of Taiwan agricultural products. However, Taiwan was having to import more than 80% of its grain.

Some places are trying to halt or at least slow the spread of human settlements onto agricultural and forested lands, and wildlife habitats. The state of Oregon, long a leader in managing growth, passed legislation in the 1970s to require urban areas to establish urban growth boundaries. This legislation was put into place to protect open space, agricultural land, forests and wildlife habitats. As a result, not only was outward expansion slowed considerably, but also many urban areas became more efficient in terms of mass transit, air quality and waste management.

Immigration to Urban Areas

Some people are pulled to urban areas in search of what they perceive to be a better life, and other are pushed to urban areas by the growth of modern, mechanized agricultural practices that favor large farm complexes over small, more labour-intensive family farms. Overall, there are three main factors that bring about immigration to urban areas:

Perceived Employment Opportunities

Much of the migration to urban areas, both in developed countries and less developed countries, is due to the increasing difficulty of earning a decent living in rural areas. The perception is that employment opportunities are greater in the cities. Most of the rural poor are farmers whose small farms or obsolete farming practices do not allow them to compete with larger, more modern, highly mechanized farming corporations.

As these rural poor lose their land, they have little alternative but to move to the cities for employment. Unfortunately, many lack the education or specialized technological training necessary to find good employment.

Usually they have no choice but to take low level jobs for very low wages and few, if any, benefits; they are often forced to work very long hours, and too often the work is in unhealthy, unsanitary and/or actually dangerous conditions. They are exposed to increasing levels of urban stress, and find themselves transformed from rural poor to urban poor. As a result, poverty is the largest single growth aspect of cities, especially those in less developed countries.

Increased Social and Cultural Opportunities

One important factor drawing people to urban areas, especially in developed countries, is the enhanced social and cultural opportunities afforded them in cities.

Cultural opportunities such as libraries, museums, orchestras, live theater and universities are located almost exclusively in urban areas. Access to these features is no doubt a factor in the movement of some people to cities. In less developed countries, and to some extent in the developed countries, educational opportunities also pull many people into urban areas.

Enhanced social opportunities exist in urban areas for some people. These include more and varied opportunities to interact with others of similar age and interest-senior citizens can interact with people their own age, young adults can meet and develop networks of friends, and younger citizens can participate in group activities such as sports and social events. In many cases, neighborhoods in inner cities actually function as small towns, and provide their inhabitants with an extended support group.

Availability of Goods and Services

There can be little argument regarding the enhanced availability of goods and services to urban dwellers as opposed to people who live in rural areas. This advantage for people who live in urban areas holds true in both developed and less developed countries, but the gap between urban and rural is greatest in developed countries. Food and water are centralized and more easily available; access to clothing is easier, as is gasoline, furniture, entertainment and bicycle parts.

Similarly, necessary services are also more readily available in urban areas. Access to health and medical care is much easier. Waste disposal is often more centralized and less of a worry to individuals; no need for a weekly trip to the local dump or landfill. In many urban areas, even those with no mass transit system, transportation is still easier and more available than in rural areas.

Urban Stresses

While there are certainly many positive reasons why people are drawn or pushed to urban areas, there are also many negative attributes to urbanization. Two important problems related to urbanization, whether in the U.S. or anywhere else in the world, are social stresses, those with a direct impact on people; and environmental stresses those with an impact on the physical environment.

Social Stresses

In both developed and less developed countries, the increased number of people, and the resulting increase in the density of those people, often bring on incredible despair. The crowding of more people into the same limited space, the poverty brought on by a too-ample supply of people willing to compete for under-paying jobs, and the fear brought on by living in areas saturated by crime are some of the social stresses to which people in urban areas are subjected.

Poverty

Many poor people from rural areas move to the city to escape poverty. Unfortunately, jobs in the cities are not as plentiful, nor is the quality of jobs nearly as high, as many urban immigrants believe. As a result, many urban immigrants are unemployed, and many others are significantly under-employed. They are transformed not from rural poverty to urban affluence, but simply from rural to urban poverty.

As a result of this poverty, the urban poor often live in abject squalor. In many cities in both developed and less developed countries, thousands or even millions of people are literally homeless, living on the streets, in parks or anywhere else they are allowed. Thousands of other people live in cardboard shanties. In the U.S., the most affluent country in the world, it is common to step around homeless people sleeping on the sidewalk within sight of the White House and the Capitol. No city in the U.S. is without its ghetto, or barrio, or slum; no city is immune. Another consequence of poverty, including urban poverty, is malnutrition. Many of the urban poor, like their rural counterparts, simply cannot afford adequate food for their families, even though a grocery market might be located in the immediate vicinity.

Compounding the effects of malnutrition, poverty prevents the poor from receiving adequate health care. They are unlikely to have even basic sanitary facilities or sources of clean water, and insect-and water-borne diseases run rampant. As a result, their quality of life is diminished, and their anticipated life span is significantly shortened.

The situation is even worse in the urban areas of less developed countries. It is estimated that in Calcutta, India, more than 70% of its approximately 12 million people live below the poverty level, and more than a half a million are homeless. The plight of the poor in Mexico City was shown to the world a few years ago when a massive earthquake killed tens of thousands of people, many of whom lived in cardboard or tin shelters.

Crowding

Crowding is a significant source of stress in many urban areas. Space is often expensive and hard to find. The result is that the majority of people is forced to live in less than optimum space and conditions. Crowding has been blamed for many social problems that seem to be the most severe in urban areas. These problems include a variety of physical, emotional and social ills, one of which is an unusually high prenatal death rate. In addition, social scientists believe that crowding is one of the factors producing the drug and alcohol abuse that seems to run rampant in crowded urban areas.

Crime Rate

Crime rates are higher in poor urban areas than in affluent urban, suburban, or rural areas. There is certainly a cause and effect relationship between poverty and crime. Teen-age ethnic gangs roam the streets with apparent impunity, establishing the local 'laws.' Anyone who tries to improve the situation is subject to killing or maiming. In recent years, senseless drive-by shootings have significantly increased. Robberies are often the result of poverty, but too often the result of the need for drugs and/or alcohol. The value of a single life is not thought to be very high.

Due in part to the crime rate in poor urban areas, a growing sense of despair is developing among the young people who live there. It is not uncommon to hear a pre-teen or young teen talk objectively of his or her likelihood of being killed before adulthood. This sense of despair leads to an attitude of "get what you can, while you can, and don't worry about the consequences."

STRESSES ON THE PHYSICAL ENVIRONMENT

In both developed and less developed countries, the increased number, and increased density of people in urban areas bring on severe stresses to the physical environment. The stresses to which the urban physical environment are subjected include effects on Air Quality, on Water Quality, on Water Resources, and on Waste Disposal.

AIR QUALITY

Nearly all urban areas have a negative effect on air quality. The problem of air pollution is discussed in detail elsewhere. Only those problems specific to urban areas or problems exacerbated by urbanization will be discussed here.

In most urban areas, automobiles are the single most important contributor of urban air pollution. In automobile-congested cities like Los Angeles, Sao Paulo, Mexico City and Bangkok, automobiles are responsible for as much as 85% of all air pollution. Incompletely burned hydrocarbons from gasoline and diesel fuel react with nitrogen oxides in the presence of sunlight to form a "brown air," or photochemical smog. Cities with effective mass transit systems, like Washington, DC and San Francisco, are less often plagued with this type of air pollution than cities more dependent on automobiles.

In the recent past, many cities that relied on the burning of coal and heavy oil for the production of energy were plagued by another form of smog. "Gray air" smog, produced by suspended particles; is still a major contributor to air pollution in many urban areas in less developed countries, including China and many eastern European countries. In developed countries where better pollution abatement procedures are followed, urban areas now have significantly less of a problem with gray air smog.

Photochemical smog is a significant problem in all urban areas, but some large cities have particularly harsh smog problems. The local climate and topography of some urban areas combine to inhibit the dispersion of photochemical smog by natural means, such as wind. Temperature or "thermal" inversions occur when weather conditions trap a layer of colder, more dense air beneath a layer of warmer, less dense air, effectively blocking the upward dissemination of pollutants. Los Angeles is severely affected by this phenomenon, and in some cases, it raises the concentration of pollutants to harmful, even lethal, levels.

Water Quality

High concentrations of people in urban areas can significantly affect the quality of the water supply. The problem of water pollution is discussed in detail elsewhere. Only those problems specific to urban areas or problems exacerbated by urbanization will be discussed here.

Urban areas produce a great deal of waste, and a high percentage of this waste is placed in sewage. A city's ability to clean and dispose of sewage is made more difficult by the great volume of water that must be handled. In most cities, the drainage from streets and buildings is combined with other types of sewage. Since most city areas are paved or covered by roofs, the majority of the rainwater that falls in a city eventually finds its way to the city's sewage treatment facilities. During times of heavy rain, these facilities are often overwhelmed and must release water without sufficient treatment. Not only does this practice lead to pollution in local lakes and streams, it can also flood areas downstream.

Water Resources

Urbanization places tremendous stresses on local water resources. Water resources for urban areas come from either surface water or ground water.

Many urban areas derive their water from adjacent rivers or lakes. Some of the surface water is diverted to local water treatment plants for purification before being used. In these cases problems arise only if the demand for water by the population being served by the treatment facility exceeds either the amount of surface water available, or exceeds the treatment capability of the water facility. It is incumbent upon urban planners to keep in mind the limits of local water resources. If these limits are exceeded, urban areas must find additional water from other resources. Urban areas that derive their water from ground water sources must always be mindful of the possibility of aquifer drawdown, which results from excessive ground water extraction. Problems arise when water demands on the aquifer exceed the rate at which the aquifer is naturally recharged. These problems are often exacerbated in coastal areas, where aquifer drawdown can lead to salt water intrusion. If the aquifer is below an urban area, the city must make certain that the water recharging the aquifer does not become polluted before or after it soaks in the ground.

Waste Disposal

Rapid population growth in urban areas also places great stresses on municipal waste disposal systems. Sewage treatment systems and municipal solid waste disposal systems, whether landfills or incinerators, have limited capacity. The problem is especially acute in areas where the soil and physiographic conditions limit the number of waste disposal sites that can be established, and thereby limit the total volume of material that can be disposed.

Emigration from Urban Areas

Over the past 40-50 years, particularly in the United States, certain populations have been moving out of urban areas. During this time there has been increasing migration from large central cities to suburbs and adjoining smaller cities. The reasons given for this emigration include:

- Escape from Urban Stresses
- Improved Infrastructure
- Establishment of New Employment Opportunities
- Perception of Better Educational Opportunities

Unfortunately, the emigration from the central city to suburban or adjacent smaller communities has a negative monetary effect on the central cities. When businesses and the more affluent segment of society move away from the central city, so do the tax revenues they generate. Budget crunches have a severe effect on the abilities of central cities to address the reasons for this emigration, and attempt to reverse it.

Escape from Urban Stresses

The single most important factor in the emigration of people from central cities to suburbs or smaller adjacent communities is to escape urban stress.

For many years, the emigration to suburbs resulted in racial and ethnic resegregation. Today, this emigration appears to produce less racial and ethnic segregation, but more socio-economic segregation. People of all races who are part of the lower socio-economic class tend to remain in the central city, and people in the middle and upper-middle socio-economic classes, whatever their race, tend to move to the suburbs. When families become economically able to do so, they move away from the crowding, poverty, crime, pollution and waste disposal problems associated with many central cities.

Improved Infrastructure

One significant reason for emigration from central cities to suburbs is that larger and more efficient streets and highways now connect the central city with outlying areas. It is easier now to get into the city for business, education or pleasure, and conversely, to return to one's living area. Some cities, most notably Washington, DC, New York City, Atlanta and San Francisco, have mass transit systems that extend into surrounding suburban areas, but such efficient systems are the exception, not the rule, among large cities. Most of the movement of commuters between central cities and suburbs takes place in automobiles-and these are most frequently occupied by only one person. Car-pooling is encouraged, but rarely required or rewarded.

Establishment of New Employment Opportunities

An important factor in the emigration from central cities to suburbs has been the location of new businesses in, and the relocation of established businesses into, the areas surrounding cities. This is especially true for smaller businesses and light industry, since they do not need to be in a city centre. In most developed countries, efficient and dependable communication and transportation facilities are readily available nearly everywhere, making it possible to do business wherever one wishes. Many business managers and CEOs feel that the less crowded, less hectic, and more aesthetically pleasing suburban or small city environment results in greater worker productivity. As businesses move, workers follow and populations grow in the suburbs and in smaller cities.

Perception of Better Educational Opportunities

A large factor driving emigration from central cities is the perception that suburban schools are better than their central city counterparts. While central city schools are not irrevocably doomed to decay and sub-standard education, there is no doubt that these schools are affected by the declining tax base that accompanies the loss of businesses and the more affluent portions of society. As available funds decline, the physical and educational quality of the schools will decline as well. What can develop is a classic case of 'the rich getting richer, and the poor getting poorer,' and the educational gap begins to widen between the 'haves' and the 'have-nots'.

Coastal Land Use

The world's coastlines are being subjected to increasing pressure as populations grow, and peoples' needs for transportation, food and recreation expand. Coasts and coastal areas serve a great variety of functions, and each must compete with the others for available space and clean water. In the United States, over 50% of the population lives within 50 miles of a coastline and more of our population is moving to the coast every year. Effective preservation of these valuable lands will have to be based on a knowledge of Coasts and Coastal Processes; an understanding of Coastal Legislation, and an appreciation of the pressures being exerted by a growing Population.

Coastal Processes

Most changes along the coast are produced by waves. In some places it is possible to see waves accomplish a significant amount of change in only a few hours. The basic process is fairly simple. Most waves are produced by storms out at sea. As the waves approach the coast, they pick up and move coastal materials. The size and orientation of the waves determines what they can do. Small waves tend to move sand and other materials onto a beach, large waves move materials off of a beach, and waves that meet the coast at an angle move materials along the beach. The amount of change that waves can bring about depends in part on the type of coastline they encounter. South Carolina's coastline is low and flat. This means that it generally receives small to moderate waves, and that changes in sea level can have a dramatic effect. Like most coastlines, South Carolina's should maintain a dynamic equilibrium, or a balance between all of these changes. As waves move sand away from a beach, there should be a matching supply of sand being added to the beach. As rising sea level moves the beach landward, the sand should be able to move with it and there should be room for the new beach to form. As people develop coastlines to create ports, resorts, and marinas, we must make sure that we do not obstruct the natural processes that maintain the beach. If we do not understand how the coastline works, it is likely that we will destroy it.

Origin of Waves

When energy from the wind is transferred to the water, waves are produced, and the waves then carry that energy as they travel for hundreds or thousands of miles. This is possible because it is energy that is moving, not water. It is easy to demonstrate this process by floating a ping-pong ball on waves in an area free of wind. The ball will move up and over each wave, and return nearly to its original location. Each wave moves the ball around in a circle, and since the ball is in contact with the water, the water must be moving in the same way. Waves represent energy passing along the surface of the water in much the same way that "waves" of energy can pass through a rug or a rope as one end is shaken up and down. In the case of waves in deep water, this circular motion is confined largely to the upper layers. Observation indicates that the waves' motion decreases with depth and nearly dies out at a distance below the waves equal to one half of their length. Thus, if waves are found to be about 40 feet long (the distance from one crest to the next), we would know that they are disturbing the water to a depth of about 20 feet. This point is known as the "wave base." Most waves in the ocean, even the very large ones, have a base no more than 40 or 50 feet deep. Since most of the oceans are several thousand feet deep, a 50-foot base is barely disturbing the surface, although such a wave would seem huge to us. This surface motion means that waves can pass fairly easily from one place to the next without a significant loss of energy. Thus the world's oceans become great collectors of energy that is steadily transported toward the coastlines. When the "waves of energy" reach the coast, they can accomplish a great deal of work by expending this energy over a small area in a short period of time.

Size and Orientation of Waves

As waves approach shallow water, they eventually reach the point at which the wave base is equal to the water's depth. As the bottom begins to interfere with the circular motion of the waves, it uses up some of their energy and begins to slow them down. When the leading waves slow down, the following ones push in behind, and the result is that the waves are squeezed together and pushed upward to become larger. As the waves become taller, they become more unstable until they fall forward and "break" on the beach. Each breaking wave can pick up a sizable load of sand, shells, or whatever is available and move it forcefully up the beach. If closely watched, this beach material will be seen to move in a number of directions:

- The most obvious motion is up onto the beach and back off again. As waves pick up sand and move up onto a beach, some of their water soaks into the beach. Since small waves contain less water, they lose a greater percentage of their volume and thus a significant amount of their ability to move sand. Larger waves hold greater volumes of water and thus lose a smaller percentage into the sand. This means that large waves can hold their loads in suspension long enough to carry the sand not only onto the beach, but also back out again. The result is that large waves tend to transport material off the beaches, while smaller waves with less transporting ability tend to build beaches up. This may be seen on many beaches that become smaller during the winter when waves are large, and grow back in summer when waves are smaller.
- The second motion is along the beach. The careful observer will note that most waves approach the shore at some angle, not straight-on. This gives the waves the ability to carry material along the beach as well as on and off of it. This means that the sand picked up by a

breaking wave will eventually be deposited some distance down the beach. As each wave repeats this process, a sizable amount of sand is moved along the beach face. This is one component of a process known as "longshore transport" that carries many millions of tons of sand and other beach material along coastlines every year.

- The third motion is the most difficult to see, but easier to experience. While playing in the surf many people realize that they are being slowly carried down the beach. This second component of longshore transport is a current that can occur in the surf zone. As waves approach the beach at an angle and break, a large amount of water is dumped into a "trough" between the beach face and adjacent rows of sand bars. The breaking waves add water to the trough, but there is no easy way for the water to escape. Seeking an outlet, the water flows along the beach carrying sand with it until it finds a gap between bars. When a gap is found the water escapes to the ocean in a strong current known as a rip. Though short and narrow, these currents account for what is often called "undertow".

Changes in Sea Level

Variations in sea level can have a dramatic effect on a coastline. Some changes in the ocean's level are small and frequent, some are very slow and gradual, and still others are rare and precipitous. The most obvious changes are the regular rise and fall of the tide. These changes are generally well known, and range from inches in some areas to 40 or 50 feet in others. Along South Carolina's open ocean beaches, the difference between high and low tide is usually between five and seven feet. Coastal residents are well aware of tidal fluctuation and plan for it, but their plans often do not include the slower changes in sea level that occur over decades and centuries. For reasons that are not well understood, the earth's climate has passed through cycles of warming and cooling for eons. When the climate begins to cool, polar ice cover increases and fresh water that would have run into the ocean is tied up as ice. The result is a drop in sea level. During the last ice age, sea level fell over 300 feet. During warming times, polar ice melts and sea level rises. Over the past 10,000 years, the earth's temperature has been increasing and the sea has been rising. During the last century the rate of warming has increased and the rise in sea level is accelerating. Evidence for this rise can be seen in ancient structures that are being flooded and coastal areas that are being invaded by the sea. There is little that humans can do to slow this rise, and thus our coastal building plans must take it into account.

Perhaps the most dangerous rises in sea level are the rapid and often unpredictable "storm surges" that accompany hurricanes and other severe storms. Storms can bring about a dramatic rise in local sea level in two ways. First of all, most large storms are associated with an area of low atmospheric pressure. As atmospheric pressure decreases, the sea bulges up just the way a balloon or a sealed bag would "bulge" as it is taken up a mountain or up in an airplane. Second of all, the effect of this bulge can be made more serious by a strong on-shore wind that "piles" the water up against the coast. The resulting surge of water can produce a "wall of water" ten to twenty feet high that can do a great deal of damage as it moves onto the coast.

WORLD ENERGY RESOURCES AND CONSUMPTION

The world has a number of energy resources which provide us with the ability to perform work (a narrowly-defined measure of the ability to effect physical change). Energy resources range from fossil fuels and nuclear fuels to renewable energy such as wind, hydro-and solar energy. When this energy is drawn upon to do work, and is converted to a less useful form, energy consumption is said to have occured.

WORLD ENERGY CONSUMPTION

In 2004 the worldwide energy consumption of the human race was estimated as 15 TW (TW=1012 Watts) by the United States Energy Information Administration. This is equivalent to 0.5 ZJ (ZJ =1021J) per year. There is at least 10% uncertainty in this number because the underlying data is continuously changing and not all of the world's economies track their energy consumption with the same rigor. Consumption by fuel type Eighty-seven percent of the world's energy is supplied by fossil fuel. The 15 TW total energy consumption of 2004 was divided as follows:

	TW=10^{12} Watt	*ZJ=10^{21}J per year*
Oil	5.6	0.18
Gas	3.5	0.11
Coal	3.8	0.12
Hydroelectric	0.9	0.03
Nuclear	0.9	0.03
Geothermal, Wind, Solar, Wood	0.2	0.006

Ever since the advent of the industrial revolution, the worldwide energy consumption has been growing steadily. In 1890 the consumption of fossil fuels roughly equaled the amount of bio mass fuel burned by households and industry. In 1900 global energy consumption equaled 0.7 TW. The twentieth century saw a rapid twenty fold increase in fossil fuels. Between 1980 and 2004 the worldwide annual growth rate has been 2%. The most significant growth of energy consumption is currently taking place in China which has been growing at 5.5% over the last 25 years. Its 1.3 billion people are currently consuming energy at a rate of 2 KW per person. The growth in energy consumption over the last quarter century. Over the last twenty years Fossil fuels have continued to grow and increased their share of the energy supply, coal has been growing the fastest. Nuclear and hydro electric energy have stagnated due to environmental concerns. With the exception of France no major western country has ordered a new nuclear power plant since 1980.

In 2004 renewable energy supplied around 7% of the world's energy consumption. Most (87%) of the renewable energy was hydroelectric power. Solar and wind power provided 4 and 65 GW (GW=109 Watt) respectively. Consumption by country energy consumption broadly tracks with gross national product, although there is a significant difference between the consumption levels of the US with 11.4 KW per person, and Japan and Germany's with 6 KW per person. In developing countries such as India the per person energy use is closer to 0.5 KW. The per capita energy consumption against the GDP per person for all countries with more than 20 million inhabitants, this covers 95% of the world population. Canada has the highest energy consumption per capita whereas the lowest energy consumption takes place in the third world.

Energy Consumption by Sector

Transportation, industry, residential and commercial (offices & shops) each consume approximately equal shares of the total 15 TW.A third of the world's energy is used to produce electricity. In 2005, global electricity consumption equaled 2 TW. The energy used to generate 2 TW of electricity is approximately 5 TW, as the efficiency of a typical existing power plant is around 38%. The new generation of gas fired plants reaches a substantially higher efficiency of 55%. Coal is the most popular fuel for the world's electricity plants. World energy resources Fossil fuel resources. Despite several voices predicting the imminent decline of fossil fuels, there are still significant reserves of all the traditional energy components. Remaining reserves of conventional fossil fuels are estimated as:

Coal	290 ZJ
Oil	57 ZJ
Gas	30 ZJ

Again significant uncertainty exists for these numbers. This uncertainty is not surprising, since the estimation for the remaining fossil fuels on the planets depends on a detailed understanding of the earth crust. This understanding is still less than perfect. While modern drilling technology makes it possible to drill wells in up to 3000 meters of water to verify the exact composition of the geology, half of the ocean has depth exceeding 3000 meters, leaving about a third of the planet beyond the reach of detailed analysis. The make up of the estimated 57 ZJ of oil on earth. The estimations for the world's oil reserves vary from low of 8 ZJ, consisting of currently proven and recoverable reserves, to a maximum of 110 ZJ consisting of available, but not necessary recoverable reserves, and including optimistic estimates for unconventional sources such as tar sands and oil shale. Both estimates provide oil for the foreseeable future at current oil consumption rate of 0.18 ZJ per year. There is a broad consensus among scientists that we are not close to running out of fossil fuels.

Coal is especially abundant and by itself can sustain the current energy consumption of the entire planet for the next 600 years. Renewable energy resources political and/or environmental considerations might move the world's energy consumption away from fossil fuels. For example, the European Commission has proposed in its Renewable Energy Roadmap, a binding target of increasing the level of renewable energy in the EU's overall mix from less than 7% today to 20% by 2020.

In 2005 the Swedish government announced their intention to become the first country to break their dependence on oil and other 'fossil raw materials' by 2020. Renewable energy sources are even larger than the traditional fossil fuels and in theory can easily supply our energy needs. The sunlight interacts with the planet. There is 89,000 TW of solar energy that falls on the planet's surface. In other words, we need to capture less than 0.02% of the available solar energy to meet our current energy needs. The available wind energy estimates range from 300 to 370 TW.

Using the lower estimate, just 5% of the available wind energy would supply the current worldwide energy needs. Most of this wind energy is available over open ocean. Not only does the ocean cover 71% of the planet, wind also tends to blow stronger over open water because there are less obstructions. The above percentages for solar and wind energy ignore the formidable challenges of energy distribution and storage that need to be solved to overcome the intermittent and seasonal variations of these energy sources. The numbers do however illustrate that it is technically possible to move away from fossil fuels should society decide to make the necessary investment. The required investment to change energy sources is enormous. It is estimated that the cost to replace the global infrastructure for liquid transportation fuels alone is $3-5 trillion.

Alternative Energy Paths

Whether a significant investment in renewable energy is wise or not, is currently the subject of much debate. Trials are taking place in several countries; Japan and Germany have started to make some investments in solar energy, making them the largest consumers of photovoltaic cells in the world despite their unfavorable geographic location.

Denmark and Germany have installed 3 GW and 17 GW of wind power respectively. In 2005 wind generated 18.5% of all the electricity in Denmark. Brazil has made a significant investment in ethanol fuel which supplies a significant part of the transportation fuel in that country. Starting in 1965, France made large investments in nuclear power and to this date three quarters of their electricity comes from their nuclear pressure water reactor plants. In the 21st century some of these different energy paths might become more mainstream and start replacing the ubiquitous fossil fuels.

EARTH'S ENERGY BUDGET

A schematic representation of the energy exchanges between the Earth's surface, the Earth's atmosphere, and outer space. The recycling of energy between the Earth's surface and the atmosphere is characteristic of the greenhouse effect. Note that the total energy entering each level is equal to the energy leaving that level as should be expected for a system in balance.

The Earth can be considered as a physical system with an energy budget that includes all gains of incoming energy and all losses of outgoing energy. The planet is approximately in equilibrium, so the sum of the gains should be approximately equal to the sum of the losses.

Note: Although the term "energy budget" is widely used, the flow of energy in and out of the Earth is actually measured in units of power (watts), not units of energy (joules). Therefore, "power budget" would be a more accurate term.

The Power Budget

The total flux of power entering the Earth's atmosphere is estimated at 174 peta watts. This consists of:

- solar radiation (99.978%, or nearly 174 petawatts; or about 340 W m^{-2})
    - This is equal to the product of the solar constant, about 1366 watts per square metre, and the area of the Earth's disc as seen from the Sun, about 1.28 × 1014 square metres, averaged over the Earth's surface, which is four times larger. The solar flux averaged over just the sunlit half of the earth's surface is about 680 W m-2
- geothermal energy (0.013%, or about 23 terawatts; or about 0.045 W m^{-2})
    - This is produced by stored heat and heat produced by radioactive decay leaking out of the Earth's interior.
- tidal energy (0.002%, or about 3 terawatts; or about 0.0059 W m-2)
    - This is produced by the interaction of the Earth's mass with the gravitational fields of other bodies such as the Moon and Sun.
- waste heat from fossil fuel consumption (about 0.007%, or about 13 terawatts; or about 0.025 W m-2).

Note that the solar constant varies (by approximately 0.1% over a solar cycle); and is not known absolutely to within better than about one watt per square metre. Hence the geothermal and tidal contributions are less than the uncertainty in the solar power.

Outgoing Power

An average albedo (reflectivity) of the Earth is about 0.3, which means that 30% of the incident solar energy is reflected back into space, while 70% is absorbed by the Earth and reradiated as infrared. The planet's albedo varies from month to month, but 0.3 is the average figure. It also varies very strongly spatially: ice sheets have a high albedo, oceans low. The contributions from geothermal and tidal power sources are so small that they are omitted from the following calculations.

The 30% reflected energy consists of:

- 6% reflected from the atmosphere
- 20% reflected from clouds
- 4% reflected from the ground (including land, water and ice)

All of the 70% absorbed energy is eventually reradiated:

- 64% by the clouds and atmosphere
- 6% by the ground

The same 70% of absorbed energy can be split this way:

- 51% absorbed by land and water, then emerging in the following ways:
  - 23% transferred back into the atmosphere as latent heat by the evaporation of water
  - 7% transferred back into the atmosphere by heated rising air
  - 6% radiated directly into space
  - 15% transferred into the atmosphere by radiation, then reradiated into space, and
- 19% absorbed by the atmosphere, including:
  - 16% reradiated back into space
  - 3% transferred to clouds, from where it is radiated back into space

Anthropogenic Modification

Emission of greenhouse gases, and other factors (such as land-use changes), modify the energy budget slightly but significantly. The IPCC provides an estimate of these forcing, insofar as they are known. The largest and best known are from the well-mixed greenhouse gases (CO_2, CH_4, halocarbons, etc.), totalling an increase in forcing of 2.4 W m-2 relative to 1750. These are less than 1% of the solar input, but contributes to the observed increase in atmospheric and oceanic temperature.

Electricity Generation

Electricity generation is the first process in the delivery of electricity to consumers. The other processes are electric power transmission and electricity distribution.

The importance of dependable electricity generation, transmission and distribution was revealed when it became apparent that electricity was useful for providing heat, light and power for human needs. Centralized power generation became possible when it was recognized that alternating current electric power lines can transport electricity at low costs across great distances by taking advantage of the ability to transform the voltage using power transformers.

Electricity has been generated for the purpose of powering human technologies for at least 120 years from various sources of energy. The first power plants were run on wood, while today we rely mainly on petroleum, natural gas, coal, hydroelectric and nuclear power and a small amount from hydrogen, solar energy, tidal harnesses, wind generators, and geothermal sources.

Electricity Demand

The demand for electricity can be met in two different ways. The primary method thus far has been for public or private utilities to construct large scale centralized projects to generate and transmit the electricity required to fuel economies. Many of these projects have caused unpleasant environmental effects such as air or radiation pollution and the flooding of large areas of land.

Distributed generation creates power on a smaller scale at locations throughout the electricity network. Often these sites generate electricity as a by-product of other industrial processes such as using gas from landfills to drive turbines.

NATURAL RESOURCES IN THE MARKET ECONOMY

The incompetence of modern economics in dealing with natural resources in their ecological totality has been voiced by many. The most penetrating description, however, comes from Georgescu Roogen who wrote:

The no deposit no return analogy benefits the businessman's view of economic life. For, if one looks only at money, all one can see is that money just passes from one hand to another: except by regrettable accident it never gets out of the economic process. Perhaps the absence of any difficulty in securing raw materials by those countries where modern economics grew and flourished was yet another reason for economists to remain blind to this crucial economic factor. Not even the wars the same nations fought for the control of the world's natural resources awoke the economists from their slumberer.

While trade and exchange of goods and services have always characterized human societies, the elevation of the market to the position of the highest organising principle of society led to the neglect of the other two vital economies in development thought. The hidden negative externalities of the development processes governed by the principles of the market have, thus, created new forms of poverty and underdevelopment. Various case studies described in this volume will substantiate such a claim. The major problem is that when exclusive attention is being given to monetary flows, requirements of natural resources not backed up by suitable purchasing power cannot be registered on the economic scene. As a result, specially in the context of Third World countries, the place of natural resources in the economy of natural resource production (or nature's economy) or in the survival economy of non-market consumption for the biological sustenance of the marginalised poor gets completely ignored.

The political economy of ecology movements cannot be understood without a clear comprehension of the place of natural resources in the three distinct economies. Ecology movements are the first indicators of compatibility and conflict among the three competing demands on natural resources. In this way the articulation of these three economies provides the foundation of a framework for an ecologically sustainable and equitable process of economic development that ensures survival and does not threaten it. The benefits and costs associated with development projects thus need to be evaluated not only in terms of the framework of the market economy but also in terms of the other two economies associated with natural resources.

THE ECONOMY OF NATURAL ECOLOGICAL PROCESSES

The terms ecology and economy are rooted in the same Greek word 'oikos' or household. Yet in the context of market-oriented development they have been rendered contradictory: 'Ecological destruction is an obvious cost for economic development'-a statement which is often repeated to ecology movements. Natural resources are produced and reproduced through a complex network of ecological processes. Production is an integral part of this economy of natural ecological processes but the concepts of production and productivity in the context of development economics have been exclusively identified with the industrial production system for the market economy. Organic productivity in forestry or agriculture has also been viewed narrowly through the production of marketable products of the total productive process.

This has resulted in vast areas of resource productivity, like the production of humus by forests, or regeneration of water resources, natural evolution of genetic products, erosional production of soil fertility from parent rocks, remaining beyond the scope of economics. Many of these productive processes are dependent on a number of ecological processes. These processes are not known fully even within the natural science disciplines and economists have to make tremendous efforts to internalize them. Paradoxically, through the resource ignorant intervention of economic development at its present scale, the whole natural resource system of our planet is under threat of a serious loss of productivity in the economy of natural processes.

At present ecology movements are the sole voice to stress the economic value of these natural processes. The market-oriented development process can destroy the economy of natural processes by over exploitation of resources or by the destruction of ecological processes that are not comprehended by economic development. And these impacts are not necessarily manifested within the period of the development projects. The positive contribution of economic growth from such development may prove totally inadequate to balance the invisible or delayed negative externalities stemming from damage to the economy of natural ecological processes. In the larger context, economic growth can thus, itself become the source of underdevelopment. The ecological destruction associated with uncontrolled exploitation of natural resources for commercial gains is a symptom of the conflict between the ways of generating material wealth in the economies of-market and the natural processes. In the words of Commoner: 'Human beings have broken out of the circle of life driven not by biological needs, but the social organisation which they have devised to 'conquer' nature: means of gaining wealth which conflict with those which govern nature."

THE SURVIVAL ECONOMY

Modern economics and the concept of development cover a miniscule portion in the history of economic production by human beings. The survival economy has given human societies the material basis of survival by deriving livelihoods directly from nature through self-provisioning mechanisms. In most Third World countries large numbers of people are deriving their sustenance in the survival economy in ways that remain invisible to market oriented development. Within the context of a limited resource base the destruction of the survival economy takes place through the diversion of natural resources from directly sustaining human existence to generating growth in the market economy. Sustenance and basic needs satisfaction is the organising principle for natural resource use in the survival economy whereas profits and capital accumulation are the organising principles for the exploitation of resources for the market. Human survival in India even today is largely dependent on the direct utilisation of common natural resources."

Ecology movements are voicing their opposition to the destruction of these vital commons so essential for human survival. Without clean water, fertile soils, and crop and plant genetic diversity economic development will become impossible. Sometimes by omission and sometimes by commission formal economic development activities have impaired the productivity of common natural resources which has enhanced the contradiction between the economy of natural processes and the survival economy.

The organising principles of economic development based on economic growth renders valueless all resources and resource processes that are not priced in the market and are not inputs to commodity production. This premise very often generates economic development programmes that divert or destroy the resource base for survival.

While the diversion of resources, like diversion of land from multipurpose community forests to monoculture plantations of industrial tree species, or the destruction of common resources, or the diversion of water from staple food crops and drinking water needs to cash crops are frequently proposed as programmes for economic development in the context of the market economy, they create economic underdevelopment in the economies of nature and survival. Ecology movements are aimed at opposing these threats to survival from market based economic development. Thus in the Third World, ecology movements are not the luxury of the rich; they are a survival imperative for the majority of people whose survival is not taken care of by the market economy but is threatened by its expansion.

The political foundation of ecology movements lies in their capacity to enlarge the spatial, temporal and social bases for the evaluation of economic development projects-in their capacity to bring into the picture all the three economies described earlier. A new economics of development will emerge only when these three economies can be conceptualized within a single framework.

TECHNOLOGY CHOICE TOWARDS HOLISTIC ECOLOGICAL CRITERIA

When economic development programmes are viewed from the perspective of all the three economies, a clearer view of the political economy of conflicts over natural resources is expected to emerge. In the dominant mode of economic development, perceived within the framework of the market economy, mediation of technology is assumed to lead to the control of larger and larger quantities of natural resources, thus turning scarcity into abundance and poverty into affluence: Technology, accordingly is viewed as the motive force for development and the vital instrument that guarantees freedom from dependence on nature ' The affluence of the industrialized west is assumed to be associated exclusively with this capacity of modern technology to generate wealth.

The concept of technology per se as a source of abundance and freedom from nature's ecological limits are based in part on the limitations of the market economy in understanding in a holistic manner, the same resources which it exploits. Only when development processes are viewed in the holistic perspective of all the three economies can the scarcities and underdevelopment associated with abundance and development be clearly seen. Most resource-intensive technologies operate in the enclaves with enormous amounts of various resources coming from diverse ecosystems which are normally far away. This long, indirect and spatially distributed process of resource transfer made possible by energy-intensive long distance transportation, leaves invisible the real material demands of the technological processes of development.

The spatial separation of resource exhaustion and the creation of products have also considerably shielded the inequality creating tendencies of modern technologies. Further, it is simply assumed that the benefits of economic development based on these modern technologies will automatically percolate to the poor and the needy and growth will ultimately take care of the problems of distributive justice. This would, of course, be the case, if growth and surplus were in a sense absolute and purchasing power existed in all socio-economic groups. None, however, is correct. Surplus is often generated at the cost of the ecological productivity of natural resources or at the cost of exhausting the capital of non-renewable resources. For the poor, the only impact of such economic activity often is the loss of their resource base for survival.

It is thus no accident that modern, efficient and 'productive' technologies 'creased within the context of growth in market economic terms are associated with heavy social and ecological costs. The resource and energy intensity of the production processes they give rise to demands ever increasing resource withdrawals from the natural ecosystems. These excessive withdrawals in the course of time disrupt essential ecological processes and result in the conversion of renewable resources into non-renewable ones. Over time, a forest provides inexhaustible supplies of water and biomass including wood, if its capital stock, diversity and hydrological stability are maintained and it. is harvested on a sustained yield basis.

The heavy and uncontrolled market demand for industrial and commercial wood, however, requires continuous over-felling of trees which destroys the regenerative capacity of the forest ecosystems and over time converts these forests into non-renewable resources. Sometimes the damage to nature's intrinsic regenerative capacity is impaired not directly by over-exploitation of a particular resource but indirectly by damage caused to other natural resources related through ecological processes. Thus under tropical monsoon conditions, over-felling of trees in catchment areas of streams and rivers not only destroys forest resources, but also stable, renewable sources of water. Resource-intensive industries do not merely disrupt essential ecological processes by their excessive demands for raw materials; they also destroy and disrupt vital ecological processes by polluting essential resources like air and water. In the words of Rothman: 'the private economic rationality of the profit seeking business enterprise is a murderous providence because it cannot guarantee the optimum use of resources for society as a whole. It cannot avoid continually creating situations which cause the pollution of an environment

In the context of resource scarcity where most resources are already being utilised for the satisfaction of survival needs, further diversion of resources to new uses is likely to threaten survival and generate conflicts between the demands of economic growth and the requirements of survival. It, therefore, becomes essential to evaluate the role of new technologies in economic development on the basis of their resource demands and conflict with the demands of survival. The productivity of 8 technology in the perspective of human survival must distinguish outputs in terms of their potential for satisfaction of vital or non-vital needs, because on the continued satisfaction of vital needs depends human survival. As Georgescu-Roegen points out.

There can be no doubt about it. Any use of the natural resources for the satisfaction of non-vital needs means a smaller quantity of life in the future. If we understand well the problem, the best use of our iron resources is to produce plows or harrows as they are needed, not Rolls Royces, not even agricultural tractors.

In the context of the market economy, the indicators of technological efficiency and productivity are totally independent of the difference between the satisfaction of basic needs and luxury requirements. between resources extracted by ecologically sensitive or insensitive technologies or of the nature of the contribution of economic growth to diverse socio-economic categories. In the context of a highly non-uniform distribution of purchasing power and scanty knowledge of or respect for ecological processes, economic growth depends on production and consumption of nonvital products. The expansion of the formal sector of the economy for the production of non-vital goods often leads to further diversion of vital natural resources. For example, water-intensive production of flowers or fruits for the lucrative export market often results in water scarcity in low rainfall areas. In a world with a limited and shrinking resource base, and in the economic framework of a market economy, non-vital luxury needs are fulfilled at the cost of vital survival needs. The high powered pull of the purchasing capacity of the rich of the world can draw out necessary resources in spite of resource scarcity and resulting conflicts.

This complete lack of recognition of the resource needs of the survival economy nature's economy in the current paradigm of development economics shrouds the political issues arising from resource transfer and ecological destruction. For the economic sector based on 'efficient modern technologies', this provides an ideological weapon for increased control of the sponsors of economic development over the total natural resource endowments of the countries concerned.

The ideological and limited concept of 'productivity' of technologies has been universalised with the consequence that all other costs of the economic process become invisible. The invisible forces which contribute to the increased 'productivity' of a modern farmer or factory worker emanate from the increased consumption of non-renewable natural resources. Lovins has described this as the amount of 'slave' labour at present at work in the world. According to him, each person on earth, on an average, possesses the equivalent of about fifty slaves, each working forty hours a week. Man's annual global energy conversion from all sources (wood, fossil fuel, hydroelectric power, nuclear) at present is approximately 8 x 10 (12) watts. This is more than twenty times the energy content of the food necessary to feed the present world population at the FAO standard per capita requirement of 3,600 cals per day.

In terms of workforce, therefore, the population of the earth is not 4 billion but about 200 billion, the important point being that about 98 per cent of them do not eat conventional food. The inequalities in the distribution of this 'slave' labour between different countries is enormous, the average inhabitant of the USA, for example, having 250 times as many 'slaves' as the 'average Nigerian'. And this, substantially is the reason for the difference in efficiency between the American and Nigerian economies: it is not due to the differences in the average 'efficiency' of the people themselves. There seems no way of discovering the relative efficiencies of Americans and Nigerians: If Americans were short of 249 of every 250'slaves' they possess, who can say how 'efficient' they would prove themselves to be.

The increase in the levels of resource consumption is taken universally as an indicator of economic development. If the present level of resource consumption in the USA is accepted as the development objectives of India, the total resource demands of 'developed' India can be calculated by multiplying the current resource consumption by a factor of 250. Neither our forests nor our fields or rivers can sustain such a 'development'. When per capita resource consumption is considered, the Malthusian argument relating population with resource scarcity does not hold good. More significant than the population factor is the total resource factor. Thus, although many countries of the South have a much larger population than those of the North, the industrialized of the world consumes more grain than all the other three-quarters put together. This high consumption is due to the fact that intensive livestock production in industrialized countries accounts for 67 per cent of their total grain consumption. This Efficient' process of livestock management for the production of meat, as reported by Odium requires 10 calories of energy input to produce one calorie of food energy.

The energy subsidy provided by the capital stock of the earth's non-renewable resources makes a resource inefficient process appear as efficient in the market economy. It is interesting to note that even in the West, nearly a century ago one calorie of food was produced by using a fraction of a calorie of energy input. The same is true in the economics of water resources use in modern agriculture. When the production of high yielding varieties of seeds is evaluated, not on productivity per unit land (tons/ha) but per unit volume of water input (tons/le lit), these miracle seeds of the Green Revolution are seen as two to three times less efficient in food production than, say, the millets. The results of evaluation of the technological efficiency of processes associated with economic development, when reexamined on a holistic basis and optimised against all resource inputs, would generally lead to the conclusion that: 'the much talked of efficiency of widely practiced high technology is not intrinsically true. They are, in fact, highly wasteful of materials and pollutive (that is, destructive to the productive potential of the environment)'. New technologies in the market economy are innovated for profit maximization and not to encourage resource prudence per se. The extent of inefficiency in the utilisation of natural resources with production processes based on resource-intensive technologies, can be illustrated with the production of soda ash, an important industrial material. In the Solvay process for the production of soda ash. the two materials used are sodium chloride and limestone.

The entire limestone used in the process ends up as waste material, 25 per cent of the sodium chloride is lost as unreacted salt. From the balance 75-80 per cent, the acidic half is lost and only the basic half goes into the final product. Therefore only 40 per cent of the raw materials consumed are actually utilised. The waste products pollute land and water resources systems. The economy of the process is artificially made good by concessions in procuring limestone, salt and fuel and further concessions in respect of land, transport, etc. It is these subsidies for natural resources which make the counter-productive processes appear efficient.

Referring to the technology of production of frozen orange juice Schnaiberg made the following remarks:

What is true of the unobtrusive shift from fresh oranges to frozen orange juice is typical of most transitions from traditional to late industrial technologies. The majority of these become more energy intensive: the energy content of all the necessary production processes increases per unit produced.... The hall mark of modern technology is its typical labour saving quality-not its energy saving aspect."

Guided by a narrow and distorted concept of efficiency and supported by all types of subsidies, technological change in market economy-oriented development continues in the direction of resource intensity, labour displacement and ecological destruction. The long-term continuation of such processes will lead to the destruction of the resource base of the survival economy and to human labour being rendered dispensable in the production processes of the market economy. The partisan assumptions of modern economic development which cannot internalise the economy of natural processes and the survival economy are thus being raised to the level of universality. As a result, with the expansion of economic development in Third World countries, the resource-intensive and socially partial development is leading to social instability and conflicts. While ecology movements in the industrially advanced countries are directed against more recent threats to survival like pollution, ecology movements in Third World countries have a much longer history related to resource exhaustion and ecological degradation of natural ecosystems. It is in these countries that the holistic ecological criteria for technology choice is needed most urgently.

The process of transformation and utilisation of natural resources for the satisfaction of societal needs determines the economic organization of human societies. At various stages of development, the dominant patterns of utilisation of natural resources have been guided by the dominant pattern of scientific knowledge, and through the generation and use of technologies that actually bridge the gap between natural resources and human needs and requirements.

A special characteristic of human societies is that they can make deliberate choices between different ways of using resources and satisfying needs. The existence of plurality of alternatives in resource use for economic development creates the need for a selection criteria to make rational decisions about the use of natural resources and technological change. A dialectical relationship exists between the criteria of technology choice and the nature of science and technology developed in response to the criteria. Traditional societies as well as modern scientific-industrial societies have adopted different systems of science and technology which differ primarily in the criteria of choice or rationality that guides resource use patterns for human needs satisfaction. The characterization of certain societies as primitive and unscientific is, thus, sociologically and epistemologically unfounded. The fact that values and rationality criteria of one form of social organization generate a particular type of science and technology matched to a particular criteria of scientificity does not imply that other social organisations lack a scientific basis for their economic activities.

If sustainable utilisation is the objective that guides the criteria of choice for a development strategy, a resource prudent technological path (T.) is rationally chosen. If maximization of the growth of man-made processes and increasing the productivity of labour is the objective, then a more resource-intensive path (T2) which is the integration of a large number of smaller technologies (t1) and in which increased resource and energy inputs allows the increase in labour productivity, is rationally chosen. In this process a large amount of secondary resources (R2-R6) are additionally required.

Traditional societies in all their diversity have, in general, shared a common set of characteristics. They have used natural resources prudently to satisfy minimum needs sustainably over centuries. Such resource use was based on.

Resource flow in resource prudent t1 and resource.-intensive t2 technology chains

1. A knowledge system with an ecological understanding of nature.
2. A technological system for processing resources to satisfy human needs with minimum resource waste.
3. Rationality criteria for demarcating vital and non-vital needs and between resource destructive and resource enhancing technologies.

Traditional world views and practices deterred over-exploitation of natural resources at all levels. As they were based on ecological perceptions of nature and guided by restraints in resource use, they used technologies which prevented ecological disruption. Modernisation of traditional societies in its present form has, by and large, been taken as synonymous with the substitution of indigenous science and technology systems by the modern western system. In this manner the resource-intensive western pattern of resource use is thrust on non-western societies through modernisation.

Modern western scientific knowledge, however, differs from indigenous knowledge systems in three important ways:

1. Modern western scientific knowledge is reductionist and fragmented.
2. Modern western technological systems are based on reductionist science and are generally more resource-intensive.
3. There are no criteria of rationality or technology choices to evaluate modern science and technology on the basis of resource use efficiency or need satisfaction capability.

These characteristics of modern western science and technology systems breaks the chain, beginning with natural resources and ending in the satisfaction of human needs and demands, into small fragments of individually identifiable economic activities. This provides justification for the resource intensity of the dominant paradigm of economic development and technological change, and thus leads to ecological instabilities.

Ecological crises are thus inevitable products of economic activities which are propelled towards longer and more complex and resource-intensive technological chains (T2) for the satisfaction of older needs (N.). Only individual segments(l) of the whole technological chain are examined from the narrow criteria of labour productivity. The situation is best exemplified in the case of food production. While indigenous and traditional food production practices used about half a calorie of energy to produce 1 calorie of food, the present mechanised and chemical farming techniques use 10 calories of energy to produce 1 calorie of food. These characteristics of contemporary scientific industrial development are the primary causes for the contemporary ecological crises. The combination of eco!ogically disruptive scientific and technological modes, and the absence of rationality criteria for evaluating scientific and technological systems in terms of resource use efficiency, has created conditions where society is increasingly propelled towards ecological instability and has no rational and organised response to arrest and curtail these destructive tendencies.

9

Changes in Forest Resources

An analysis of changes in the cover and condition of the world's forests requires a differentiation between:

1. the increase in forest cover (by afforestation or natural colonization of trees on non-forest land) or decrease (by deforestation) of forest area; and
2. changes in forest condition, either positive (recovery of degraded stands, stand improvement treatments) or negative (decline, defoliation or dieback, effects of forest fires, degradation through unsustainable exploitation for wood, overgrazing, effects of pests and diseases).

CHANGES IN FOREST COVER

Between 1980 and 1995, the extent of the world's forests decreased by some 180 million ha, an area about the size of Indonesia or Mexico. This represents a global annual loss of 12 million ha, an area equivalent to the size of Greece or Bangladesh. During this 15-year period, developing countries lost nearly 200 million ha of natural forests, mostly through clearing for agriculture (shifting cultivation, other forms of subsistence agriculture, the establishment of cash crop plantations such as oil palm, and ranching). This was only very partially compensated for by the establishment of new forest plantations. Over the same period, forests in the developed world expanded slowly (by some 20 million ha) through afforestation and reforestation, including natural regrowth on land abandoned by agriculture.

Forest plantation and natural regeneration on abandoned agricultural land more than compensated for clearing of forests due to urbanization and infrastructure development in most industrialized countries, outside the former USSR. While the gain of forest cover in developed countries was 20 million ha in the period 1980-95, 9 million ha of this increase occurred in the 5 years from 1990 to 1995. The situation is quite different in the developing world, where deforestation exceeded net afforestation/reforestation, particularly in the tropical zone. As a whole, the annual rate of deforestation in the developing world between 1990 and 1995 was 0.7%, equivalent to 12.6 million ha, with the highest rate in tropical Asia-Oceania, closely correlated with population and income growth. The lowland forest formations were the most affected by deforestation, although the proportional loss during this period was greater in upland formations (1.1% compared with 0.8% in the lowland formations). There is some evidence that the rate of loss of forest cover in developing countries was slowing towards the end of the 15-year period 1980-95. The annual rate of forest loss in the period 1980-90 was 15.5 million ha compared with 13.7 million ha between 1990 and 1995.

CONVERSION OF FORESTS TO OTHER LAND COVER

The Forest Resource Assessment 1990 carried out an assessment of the relative importance of the various factors involved in deforestation at regional and global levels between 1980 and 1990 in the whole tropical belt. Among the most significant outputs of the study were the 'area transition matrices' of the type, which indicate transfers from one land cover class to another over the 3068 million ha of the tropical zone covered by the sample. For instance, the first row shows that 1275.9 million ha of the total area of 1368 million ha of closed forest in 1980 remained as closed forest up to 1990 and also shows the fate of the 92.1 million ha converted to other land cover classes (open forest; long fallow; fragmented forest; shrubs and short fallow; other, *i.e.* non-wooded, land cover; and forestry or woody forest and agricultural tree plantations). All these transfers involved changes of woody biomass attempts to capture by replacing each class along the *y*-axis (average biomass per hectare) and showing the importance of transfers along the *x*-axis.

Changes in Forest Condition

Factors affecting the health and vitality of all types of forests have attracted increasing attention in recent years. Outbreaks of fires have made headline news in developed and developing countries alike, while insect and disease attack and generalized declines in forest condition have caused more localized concern. Insects and disease and forest fires exist under 'natural' systems and conditions; what has caused concern has been their spread in artificial systems and in conditions considerably modified by humankind, resulting in direct economic loss.

Fire arising from natural causes has been a major influence and is a driving evolutionary force in several forest ecosystems, such as the pine forests of Central America. Fires are important to the health and maintenance of these ecosystems and the total exclusion of fire from them can lead to the build-up of fuel and ultimately abnormally destructive fires that cause the loss of the ecosystem, as happened with the fire exclusion policy of the National Parks Service of the USA and the disastrous fires of 1988. However, many fires arising from human activities, whether planned or unplanned, have damaged and destroyed large areas of forests and woodlands. Resource managers face a demanding public often with conflicting needs and incomplete information, leading to the development of new policies that take into account public attitudes towards fire management.

At the global level, an agreement was signed at a meeting in 1990 to initiate the International Decade for Natural Disaster Reduction and the target date of 2000 was set for signatory countries to effectively monitor and manage wildfires in their respective countries. In view of the lack of information on causes, extent and the number of fires at present, and the difficulties in collecting data, this target date is unlikely to be met.

Developed Countries

Total forest area is slowly increasing in the developed world, although the need for protection from fire, insects and disease has continued and some aspects of forest condition give cause for concern. Although the widespread decline and death of certain European forests due to air pollution predicted by many in the 1980s did not occur, deteriorating forest condition has remained a serious concern in Europe and North America. The main causes of the decline of forest condition in Europe have been low soil moisture availability due to drought, and high temperatures. These factors, combined with airborne pollution and the use of non-adapted seed sources in forest plantation development, have predisposed forests to attack by insects and disease and to decline. Forest damage due to air pollution is severe in some parts of central and eastern Europe in particular. Air pollution is also said to have a role in the decline observed in certain species in other industrialized countries, such as sugar maple *(Acer saccharum)* in eastern Canada, the high-elevation forests of red spruce *(Picea)* in eastern USA and *Cryptomeria japonica* stands in Japan.

The exclusion of fire may have caused the decline of some forest ecosystems, such as oak stands in central USA or natural eucalypt forests in Australia where fire is an integral part of the ecology of some plant communities and here it may be used to facilitate their natural regeneration.

Despite an overall increase in the number of fires (by some 40% in Europe, 8% in North America and 120% in the former USSR), wildfires over large areas of forest have become less frequent in the industrialized countries. Further, the average area burned by wildfire decreased in most regions between 1983-90 and 1991-94. There has been a slight reduction in total areas burned in the developed world overall as a result of improved prevention, detection and control systems, although there has been an increase of 15% in the former USSR. In 1990, the average area of forest and other wooded land affected annually by fires in Europe, North America and the former USSR together was about 4.26 million ha or 0.22% of their total area of forest and other wooded land. Pests and diseases remain constant threats, particularly to the seminatural forests and plantations in the industrialized world. Trees are more likely to be attacked when under stress, whether from abnormal weather conditions such as drought or high temperatures, airborne pollution, uncontrolled movement of germplasm or lack of management. Recent outbreaks of pests and diseases that may have been related to stress include the decline of various species of oaks in many parts of Europe and in central Russia, the recurrent infestation of forests in Poland by the nun moth, the attack on beech by the beech scale in western and central Europe, birch and ash diebacks in north-eastern USA, the littleleaf disease of shortleaf pine in southern USA, and the 'x-disease' of pines in southern California.

FOREST CONDITION IN DEVELOPING COUNTRIES

Less attention has been paid to forest condition in developing countries because reduction in forest area has attracted more attention. This reduction in density has often been the result of over-exploitation for timber and fuelwood. Overgrazing and repeated bush fires are other significant causes of decrease in density, especially in the dry tropical and non-tropical zones.

Every year, very large areas of savannah woodland and mixed forest-grassland formations are affected by fires set by herders to provide an early flush of green grass when the rains start, particularly in the dry zones of Africa and South America, although no reliable data are available on their extent.

Forests in the humid tropics have also, at times, been affected by large fires, the most serious in recent years having been those associated with the dry weather conditions arising from the El Niño Southern Oscillation (ENSO). In 1997-98, these caused large-scale destruction of logged and secondary forest in Indonesia (particularly Kalimantan, Sumatra and Irian Jaya), Mexico and Central American countries, as well as leading to smoke pollution over wide areas beyond the borders of these countries. The cause of these fires has been burning for land clearance, but it should be noted that the fires are not necessarily in forest; although described as 'forest fires' they are frequently in grassland or on land that has been cleared of forest. It is *estimated* that of the 2 million ha burned in Indonesia in 1997, 150000-200 000 ha may have been in forest. Other serious fires associated with ENSO effects have included that in East Kalimantan, which burned 3.6 million ha in 1983.

Coniferous forests in the humid tropics have often been affected by fires, although it must be noted that fire is frequently necessary to maintain and regenerate them. In the 1980s, the area of pine forest in Honduras and Nicaragua burned annually amounted to some 65 000ha (or about 3.5% of the total pine forest area of these two countries), and widespread fires in natural and artificial tropical pine forests occur in other countries such as Guatemala, Mexico and Indonesia (northern Sumatra).

In the subtemperate and temperate zones of the developing world, fire is also a permanent threat to forests, particularly when they are no longer used by local people for grazing and other purposes. In the 1980s, the average area of forest and other wooded land burned annually was 140 000 ha in the temperate/ subtemperate zones of South America (including southern Brazil). From 1950 to 1990, fires in China are reported to have affected an average of 890 000 ha annually, the most damaging one having been the 'May 6' fire, which burned some 1.85 million ha in the north-eastern province of Heilongjiang in 1987.

In the absence of a global statistical fire database, it is difficult to provide an overall estimate of the annual extent of fires in forests and other wooded lands.

A very crude estimate for the temperate/subtemperate and humid tropical zones of the developing world (leaving aside the significant dry tropical zone, for which little reliable information exists) would be of the order of 2 million ha of forest and other wooded land annually during the 1980s. Given the lack of sufficient capacity in fire prevention and control in most developing countries, no significant reduction in wooded areas burned is likely to occur in the near future.

Outbreaks of pests and diseases in developing countries are generally reported for those plantations and planted trees where the impact is most apparent. Introduced pests are usually extremely destructive and several of them have had very damaging effects in recent years in the developing world. Examples of introduced insects affecting plantations and planted trees include the *Leu-caena* psyllid, *Heteropsylla cubana,* which has spread into Asia and the Pacific islands and is now extending across Africa; the cypress aphid, *Cinara cupressi,* which is established in eight eastern and southern African countries and is causing heavy mortality among a number of exotic and indigenous species but especially the important plantation species *Cupressus lusitanica,* threatening its future role in the plantation programmes of these countries; and the European woodwasp, *Sirex noctilio,* which has spread into Argentina, Uruguay and southern Brazil, affecting principally *Pinus taeda* but which may become a threat for the large Chilean plantation estate of *Pinus radiata.*

Strategies to control introduced insects from attacking forest trees include regulatory measures, eradication, integrated pest management, and monitoring of population levels and occurrence of potentially harmful species.

Effective monitoring of insect populations will be especially important in view of the large plantation programmes being carried out to produce industrial roundwood, which often rely on one or a few exotic species. Such programmes may be able to afford the investment that is necessary for control measures. Smaller-scale programmes may have to insure against failure due to unexpected attack by insects or diseases by diversifying into several species adapted to the sites and end-uses.

There is also evidence of forest decline due to a combination of biotic and abiotic factors in the developing world, with air pollution likely to play an increasing role in some cases as industrial and transportation infrastructure develops.

Examples of such decline include neem *(Azadirachta indica)* in the Sahel, framiré *(Terminalia ivoren-sis)* in Ivory Coast and Ghana, and *Eucalyptus globulus* plantations in Colombia and Peru, while airborne pollution is affecting forests near large cities and industrial areas in China. Despite the overall significance of the corresponding damage and losses, surveys of forest decline and diebacks in developing countries remain all too rare.

HISTORICAL PERSPECTIVE OF FORESTS COVERED

Forests covered about half of the earth's surface up to the development of early civilizations but today cover less than one-third of that area. Massive forest clearances were a feature of many early cultures, including ancient Assyria, Babylon, China, Egypt, Greece and Rome. Much of this forest clearance was to provide land for agriculture, although these societies were also voracious consumers of wood for cooking, heating, copper smelting, pottery making, brick-firing, house construction and shipbuilding, and this led to deforestation wherever the forest did not have the opportunity to regenerate. For more than 10000 years the forests of the Mediterranean region have been cleared; they now cover about one-sixth of the region and those that remain, often on land that cannot readily be cultivated, have been degraded by humans and their animals through unmanaged grazing. The export of high-quality cedar wood from Lebanon to Egypt began nearly 5000 years ago and led, eventually, to the virtually complete destruction of the cedar forests. The clearance of forests is thus not a new phenomenon in human history, except in its present scale and spread.

At times, the pressure on the forests eased. The decline and ultimate fall of the Roman Empire following the sack of Rome by the Goths in the fifth century AD, and the subsequent economic decline of much of Europe, led to the regrowth of forests in some areas. The same happened at the time of the Black Death in the mid-fourteenth century when the population of western Europe was reduced by over one-third, and in some places by half. In Germany, people fled rural life and the oppression of rural overlords into the depopulated cities, while many responded to the invitation of the Polish kings and moved to the eastern territories (the Baltic, East and West Prussia, Pomerania). As a result, the forests of central and western Europe colonized and regenerated over large areas, and it was not until the sixteenth century that population pressure and the demand for forest products again increased to a level comparable to that at the end of the thirteenth century.

Following the Thirty Years War (1618-48), famine and disease again reduced the population of rural Germany, on this occasion by 40%. This pattern was repeated during the Spanish Heritage War (1701-14) and the Seven Years

War (1756-63). Following the order of the French King Louis XIV, nearly all towns and castles in south-western Germany were burnt. Their reconstruction consumed an enormous amount of timber so that by the middle of the eighteenth century much of the forest wealth of central Europe had been destroyed, a timber shortage ensued and there was a loss of soil fertility and soil erosion. Vast tracts in northern Germany were degraded to deserts with shifting dunes.

In Britain, where population growth, land clearance and economic development were especially rapid, the natural forests had been heavily depleted by the late Middle Ages. Iron smelting, relying on charcoal, and other wood-based energy users encountered supply difficulties. By the sixteenth century most of the accessible wood resources had been exhausted in England and iron-makers turned to the woodlands of Scotland, Ireland and Wales. Timber began to be imported into England in large quantities for every purpose.

Destruction of the natural forests has thus taken place over most of Europe. It accelerated as the demand for fuelwood and timber grew rapidly in the early years of the Industrial Revolution, but with the continuation of economic growth and industrialization the pressures on the forests have gradually eased. By the early years of the present century, the total area of the European forests had more or less stabilized and in the past decade has even increased slightly. In particular, the pressure on the Mediterranean forests for the supply of fuelwood and land for agriculture and grazing has declined, although the risk of damage from fires continues.

A similar process of forest clearance occurred in the USA. At the time of European settlement in the early seventeenth century about 46% of the land was forested; by 1992 this area had declined to about one-third. Huge areas of forest were cleared in the eighteenth and nineteenth centuries as the population grew and industrialization progressed. Early clearing was for agriculture; in 1800, 95% of the people lived on the land, mostly practising subsistence farming. Most wood was used for firewood even in the late nineteenth century, which depleted forest resources especially around population centres. Fencing was the next most important use of wood, followed by early industrial use. The spread of the railroad then further increased demand for wood, which was used for locomotive fuel until the mid-eighteenth century but above all for sleepers or railroad ties. For example, in 1900 it was estimated that the replacement of ties on existing track required the output of 6-8 million ha of forest annually.

The importance of the sea for defence and trade led to laws being passed in several European countries in an effort to protect and expand wood resources for naval vessels and merchant fleets, especially oak for building the hulls and pine for masts. The laws were only partially successful, even for these limited objectives. The forest cover of France had shrunk from perhaps 35% of the country's area at the beginning of the sixteenth century to about 25% by the middle of the seventeenth century, and despite the reforms of the administration of the forests and of the forest laws by Louis XIV and his minister of finance, Colbert, in 1661, the situation continued to deteriorate due to demands for shipbuilding and the smelting of iron ore. Nevertheless these reforms began a forestry tradition and laid the basis for a strong forest service.

A similar realization that the forest resource was not limitless led to the emergence of rules to control the use of forests in Germany, Denmark and The Netherlands in the eighteenth century, in India in the mid-nineteenth century, in Sweden and Finland in the late nineteenth century and in the USA in the early twentieth century. At that time colonial powers were reserving forests, introducing forest laws and establishing forest services in East Africa, Malaysia and many other now-independent countries. In the eighteenth century, efforts to restore forest cover and guarantee future timber supplies were made. Today Germany has 30% forest cover, which originated in the plantations of the seventeenth century; initially these stands were largely of single species (oak, beech, Scots pine and Norway spruce) but today are slowly being converted into mixed stands with uneven age structure.

The development of management techniques for natural forest began in Europe for a number of purposes. Sometimes this process was accelerated for specific objectives, for example to reduce the risk of avalanches, to control erosion, to stabilize sand dunes or to protect agricultural crops, as well as to provide firewood for towns. Management planning in the sense of regulation of yield had been introduced in France by Colbert, while in Denmark management plans with rotations of oak of 100 years were instituted in 1763.

Originally recruited from retired soldiers, the foresters of the specialized forest services created by the rulers of various German states were to have a profound influence on forest science and practice even up to the present. The experience they gained, combined with traditional forestry knowledge, was published and widely disseminated. One of the best-known publications of that time was *Sylviculture! Oeconomica,* published in 1713 by Hans von Carlowitz (1645-1714). Von Carlowitz was a mining engineer in Saxony preoccupied with the insufficient timber supply for the Saxon mines. His book dealt not only with timber harvesting, species selection, utilization of timber and sowing of conifer seeds to reforest clear-fellings, but also with promoting the introduction of foreign (exotic) species and he was the one who first coined the term *Nachhaltigkeit* (sustainable forest management). The handful of German foresters who became the founders of forest science included Heinrich Cotta (1763-1844), Wilhelm Pfeil (1783-1859), Johann Christian Hundshagen (1783-1834), Carl Heyer (1797-1856), Gottlob König (1779-1849) and George Ludwig Hartig (1764-1837), who as professor at the University of Berlin introduced forestry training into the Prussian forest service and had a strong impact on the forest management principles of his day, not only in Prussia but all over central Europe. Hartig's rule was that no more nor less timber should be extracted annually from state forests than could guarantee perpetual supply, which became the principle of central European forestry, *i.e.* sustainable forest management.

The reduction in forest area had been halted or even reversed in the industrialized countries by the 1920s. Since then the standing volume of forests has increased with better management and increased protection, and in Europe the yearly forest harvest of many countries is smaller than the annual increment. The reasons for this historical transition from rapidly diminishing to stable, or even increasing, forest areas in many countries of the Northern Hemisphere with industrial or service economies are complex. The substitution of coal for fuelwood for industrial and domestic use was one obvious factor. However, more important were the major demographic and economic changes that took place as industrialization and economic growth proceeded.

Rural populations fell drastically as people migrated to the cities and towns. By the turn of the present century, the average proportion of the population living in the rural areas of the industrial countries was around 40% compared with 80% or more at the beginning of the Industrial Revolution. Today, in Europe and the USA, the rural population is about 25% of the total. National population growth rates have dropped to around replacement level in most of the industrial countries. Equally important were the changes that took place in agricultural production. Subsistence farming gave way to an increasingly capital-intensive form of agriculture relying on mechanization and high inputs of fertilizers and pesticides and, in animal husbandry, the stall-feeding of animals. Rather than requiring ever-expanding amounts of land, agriculture became more productive, leading eventually to the present politically charged problems of crop surpluses and the need to 'set aside' or take increasing amounts of farmland out of food production. Additionally, increased public awareness of the value of forests and of environmental issues in urbanized societies of the postindustrial or service economies, combined with increasing demand for new services from forests such as recreation and amenity, have been important recent trends in stabilizing forest areas.

Although history does not repeat itself exactly, the experience of the industrial world helps shed light on what is happening in many developing countries in relation to the pressures on their forests. Expanding populations (especially those existing at subsistence level), accompanied by significant rural-urban migration, result in increased demand for agricultural production. Even highly punitive forest protection legislation has been largely ineffectual in the face of the desperation of people trying to grow enough food, or graze sufficient animals, to survive.

It is only with increased agricultural production supplying stabilized populations that the pressure to clear forest land diminishes. Indeed, sustainable forestry may be possible only where agriculture is, if not sustainable, at least stable. This transition can be seen taking place today in some of the rapidly industrializing countries of the developing world. The Republic of Korea, for example, where the rural population has fallen sharply while agricultural production has increased dramatically, has reached the turning point at which the area of forest can begin to expand again, aided by massive reforestation programmes. In others, such as the newly industrializing countries of the Pacific Rim, forest depletion continues, although the economic and demographic conditions and social demands for stabilizing the remaining areas of forest are beginning to emerge.

In many other developing countries, however, rural population growth continues virtually unabated, agricultural productivity is low and economic stagnation means that there are few employment opportunities for those wishing to leave the land. In such cases, increasing pressure to clear forest for agriculture is inevitable whether legalized by land reform or not.

The parallels with the experience of the industrial world are thus obvious but what makes today's position in the developing world different is the sheer impact of population numbers, the pace of change and the fragile nature of many tropical soils and their inability to be used for long-term agriculture. Regarding population numbers, for example, the population of the European continent grew from about 140 million in 1750 to 265 million by 1850, 392 million by 1950 and 499 million by 1990. The population of the USA grew from 23 million in 1850 to 151 million in 1950 and 248 million in 1990. Today, on the other hand, around 50 million people are being added to the rural populations of the developing world *every year.* Where there were about 12 ha of forest per head of the global population of about 500 million people in 1750 there is now about 0.75 ha per head, and the rate at which the forests are disappearing, especially tropical forests, is far faster than at any other time in human history, despite net gains of forest cover in some developed countries and some evidence of a slightly slower rate of deforestation in developing countries in recent years.

RECENT ESTIMATES OF GLOBAL FOREST AREA

The following section is based on 1990 baseline figures, updated to 1995, prepared by the Food and Agriculture Organization (FAO) Forest Resource Assessment Programme, reported in and largely drawn from *State of the World's Forests 1997*.

In 1995 forests were estimated to cover 3454 million ha, or 26.6% of the total land area of the world (Greenland and Antarctic excepted). Almost two-thirds of the world's forests were located in seven countries: Russia, Brazil, Canada, USA, China, Indonesia and Zaire; 29 countries had more than half of their land covered by forest, of which 21 were in the tropical belt. However 49 countries, in addition to the many non-forested small island states and territories, had less than 10% of their land covered by forests. Five entire subregions were in this category: North Africa (1.2% of the land area), Near East (1.9%), Temperate Oceania (6.2%), Non-tropical Southern Africa (6.8%) and West Sahelian Africa (7.5%).

The latest information on the distribution of forest and of forest cover change by ecological zones was provided in 1990; the results of the Forest Resources Assessment 2000 were released by FAO in 2000. In 1990, temperate and boreal forests occupied 1.64 billion ha and tropical forests 1.76 billion ha. A breakdown by ecological zone was available only for tropical forests. The figures on tropical forests show that most (88%) are in lowlands; of these, tropical rain forests accounted for 47% of all lowland tropical forests, followed by moist deciduous forest (38%) and dry and very dry formations (15%).

NATURAL FORESTS

The forest area estimates described above include undisturbed forests, forests modified by humans through use and management (or 'seminatural' forests) and forests created artificially by humankind (*i.e.* forest plantations) by afforestation or reforestation. (Afforestation is defined as the establishment of a tree crop on an area from which it has always, or for a very long time, been absent. Reforestation is defined as the establishment of a tree crop on forest land.) In most industrialized countries, particularly in continental Europe, forests are being managed in such a way that at management-unit level a continuum exists from low-intensity management, involving natural regeneration, through more intensive methods involving some artificial planting to highly intensive methods with complete planting and cultivation; this makes it difficult to isolate figures for natural forest and plantations. The distinction between natural or seminatural forests and forest plantations can more easily be made for developing countries and some industrialized countries such as New Zealand in which forest plantations have been established using introduced species.

Interest in natural forests, particularly their role in the conservation of biological diversity, has led to efforts to compare forests today with what is thought to be their original character and to give complete protection to areas of forests which have had no, or minimal, human interference. Although there are difficulties in identifying the extent of natural forest, compounded by problems of definition, some information exists that can be used as an indication of broad patterns of natural forests in various regions.

An attempt has been made by the World Wide Fund for Nature (WWF) to quantify the area of forests in western Europe that has been relatively undisturbed by humans or which has retained much of its natural character. Their report distinguishes between 'virgin forest', defined as 'forest ecosystems whose characteristics are determined exclusively by natural location and environmental factors... without human influences present or visible any more', and 'natural and ancient seminatural forests', which 'have not been planted or sown by man for the past two centuries' and 'which continue to have a large number of the natural elements'. The study found that only a small proportion (probably < 1%) of the total forest land in northern and western Europe could be considered as virgin forest, which has arisen since the last glaciation. Almost all was located in Sweden, Finland and Norway, with small areas in Greece, Austria and Switzerland and (according to another author) in France. In eastern Europe, Slovakia and Belarus have considerable areas of virgin forest while Poland and Croatia have small areas. In addition, in northern and western Europe, the WWF report identifies natural and ancient seminatural forests representing 2.1% of the total forest cover (1990) of the 16 countries concerned. There are a further 3 million ha of land in the region in national parks and other protected areas and another 50 000 ha in small forest reserves (mostly for nature conservation and scientific research), whose use is tightly restricted.

The situation in temperate and boreal North America is quite different from that of densely populated Europe and Japan, where use and management of forests for many centuries have left very little of the original forest area untouched. 'Old growth forests', as they are called in North America, still cover extensive areas. On the lands managed by the US National Park Service alone, old-growth forests covered 1.97 million ha in 1988. Although not a direct measure of the extent of old growth forest, it may be noted that the area of forest included in national parks and other protected areas in North America (USA and Canada) was reported to be nearly 49 million ha in 1990.

The FAO Forest Resource Assessment 1990 did not distinguish between undisturbed and disturbed 'natural' forests in developing countries. However, the Forest Resource Assessment 1980 made estimates of the areas of undisturbed closed forests (primary forests and old secondary forests where there had been no logging for the last 60-80 years) and of closed forests that were included in national parks and other protected areas (thus relatively undisturbed, at least in theory). At that time, these two categories together represented 60% of the total closed forest area in the tropics, a proportion varying from 39% in tropical Asia to 59% in tropical Africa and 69% in tropical America. These different proportions by region reflected a slower development of large-scale harvesting in tropical America compared with tropical Africa and Asia, and also the fact that, in tropical America, spontaneous colonization did not follow in the wake of logging as systematically as in the two other regions due to lower population pressure. Although the two categories do not match the concept of 'virgin forests' of Europe and of 'old growth forests' of North America discussed above, the sum of the two categories nevertheless gives an indication of the amount of forest disturbance (or management, depending on the point of view of the observer) that existed around 1980 in the humid tropics. Although corresponding estimates for 1990 and 1995 are not available, it is likely that the share of undisturbed forests remains higher in the three tropical regions than in Europe and probably in North America too.

Forest Plantations

Recent data in this chapter are also available in *State of the World's Forests 1999*, which also discusses some of the issues concerned with forest plantations in more depth.

Early Development

By the seventeenth century the decline in the area of native forest in European countries led to the planting of trees, largely to provide alternative sources of timber supply. Sometimes plantations were established for the provision of services or other products than timber. They might be planted as shelterbelts, for dune stabilization, for amenity or for the supply of firewood.

Plantations of native species were established at first in areas where forests occurred naturally, for example France, Germany, England and Scotland, but later also on previously unforested land. In France, for example, planting of *Pinus pinaster* was started on the sand dunes of the Landes and with *Pinus sylvestris* and *Picea abies* on former agricultural land in the Vosges; in Germany likewise agricultural land was planted with Norway spruce *(Picea abies)* in Saxony. In tropical countries such as Myanmar (then Burma), teak *(Tectona grandis)* was planted as a native species in the *taungya* system.

Teak was one of the first exotic forest plantation species to be used, being planted in Sri Lanka and the island of Java (Indonesia) from early in the nineteenth century. Exotic species, such as Douglas fir *(Pseudotsuga menziesii)*, Sitka spruce *(Picea sitchensis)*, Japanese larch *(Larix kaempferi)*, lodgepole pine *(Pinus contorta)* and poplar *(Populus deltoides)*, were introduced to Europe during the nineteenth century, when they began to play an increasingly important role in forest plantation programmes. Other exotic species, particularly the eucalypts *(Eucalyptus* spp.) and wattle *(Acacia* spp.) from Australia, were introduced as exotics in tropical and subtropical countries from the middle of the nineteenth century, while *Pinus radiata* was introduced from California to New Zealand and other countries such as Chile from the early years of the twentieth century. Box 1.3 describes in detail the experience with forest plantations in Denmark.

The main reason for the establishment of plantations remained the decline in natural forest area and a scarcity of wood, the conversion of woods and forests to agricultural and grazing use. In recent years increased areas of natural forest have been managed for nature conservation, recreation, wildlife parks, etc., and commercial wood production has been either reduced or eliminated. For example, over 25 000 ha of high-yielding hybrid poplar plantations have been established in the north-western USA between 1992 and 1997 in response to both increased demand for poplar wood for orientated-strand board and decreased supply from public forests. On the other hand, some new areas have become available in European countries as land is taken out of agriculture due to trade and market considerations.

Objectives

Forest plantations are tree crops that are in some, but not all, ways analogous to agricultural crops. They often have a simple structure, at least in youth, and are usually composed of one or a few species (but not varieties as in agriculture, except in a few cases such as the intensively bred poplars) chosen for their fast growth, yield of specified products and ease of management. When established for wood production they have a higher productivity of usable wood than natural forests, but due to the way they are managed they do not, indeed cannot, provide the full range of goods and services that natural, seminatural or even secondary forest can provide.

Although the objective of many plantations is the production of industrial roundwood and/or fuelwood, many are established for environmental protection or other services and some also provide non-wood forest products, such as fodder, various foodstuffs, medicines, etc.

Area

The area of forest plantations throughout the world started to increase in the 1970s as many governments became concerned about wood supplies for industry, and in developing countries fuelwood supplies, and has continued to increase since. However, there are no reliable global figures for plantation areas because forests of native species in several developed countries in the temperate and boreal regions, especially in continental Europe, are frequently regenerated naturally and it is not possible to distinguish those areas where supplementary artificial planting has been done. Furthermore many countries consider their plantations as 'seminatural' forests over a certain age, because as stands mature the clear initial row layout of the trees is lost and other species naturally regenerate under the canopy. For example, Austria, Czech Republic and Finland, in responding to a questionnaire to collect data for the ECE/FAO Temperate and Boreal Forest Resource Assessment (TBFRA), stated that they had no plantations in their countries as defined by the TBFRA process.

In 1995, an approximate estimate of the area of plantations in developed countries was 60 million ha, comprising 13.7 million ha in North America, 22.2 million ha in the Commonwealth of Independent States, 12.1 million ha in Europe and 13.2 million ha in Oceania (Australia, New Zealand, Japan). The most significant areas of plantations were in the Russian Federation (17.3 million ha, 2.1% of the country's total forest area), the USA (13.7 million ha, 6.3%), Japan (10.7 million ha, 44.4%), Ukraine (4.4 million ha, 46.8%) (FAO/

ECE 1996), Spain (1.9 million ha, 14.9%), New Zealand (1.54 million ha, 19% of the total forest area in 1996, 91% of which was *Pinus radiata)* and Australia (1.04 million ha in 1994, 85% of which was softwood species, mainly *P. radiata).*

The estimated 'net' plantation area of 55 million ha in developing countries in 1995 was about 2.8% of the total area of forests in developing countries. In 1980 the net plantation area was assessed at about 40 million ha. It has thus increased by about 15 million ha in 15 years, but even this figure may be liable to error. Many developing countries, especially those with large forest plantation programmes, provided updated information on their present plantation plans to FAO in 1996 and 1997 from which the reported annual rate of new plantations of 3 million ha was derived. Note, however, that this is the reported or planned rate, which may not necessarily have been achieved. Most of the countries with large plantation estates indicated that they intended to double their plantation areas between 1995 and 2010.

It was estimated from reported figures that 57% of the forest plantation area consisted of hardwood species and 63% was established for industrial purposes. Nearly three-quarters of these plantations were in the Asia-Pacific region, where China (21 million ha) and India (20 million ha) dominate, while about 15% were in Latin America and 10% in Africa.

One of the trends in the tropics has been that the proportion of industrial plantations established in large blocks fell from 40% of the total plantation area in 1980 to 35% in 1990. The proportion of smaller plantations established through farm forestry or agroforestry programmes grew in importance during the period 1980-95, particularly in the Asia-Oceania region. Unfortunately, the figures on small-scale private or community-owned forest plantations are even less reliable than for large-scale plantations. Some of these farm forestry or agroforestry plantations supply industrial wood markets for pulpwood. In many countries, particularly those with limited forest area, planted trees grown outside the formal forest area often provide the bulk of fuelwood, poles, construction wood, utility wood, as well as fodder and other non-wood forest products for household use.

Of the area of hardwood plantations planted for industrial use, 30% or nearly 10 million ha consists of eucalyptus, followed by acacias (3.9 million ha or about 12% of the hardwood area) and teak (about 7%). Short rotation plantations of hardwood species such as the eucalypts, acacias and *Gmelina arborea* have been grown for many years by the private sector, but the establishment of plantations of teak or other valuable hardwood species has been carried out only by government forest services because of their slow growth and hence delayed returns. However, the likelihood of reduced supplies of high-quality hardwood logs derived from natural forests, combined with increasing purchasing power and expected higher prices for logs, is leading to increasing interest in investment by the private sector in valuable hardwood species, especially teak, in a number of countries, for example India,

Malaysia, Costa Rica and Ghana. Of the softwoods grown for industrial purposes, fast-growing pines such as *Pinus radiata, P. patala and P. caribaea* constitute about 25% of the area while other, often slow-growing pines make up about 36%.

Responsibility for the monitoring and regulation of plantation crops grown for food and certain other purposes has long been the job of the agricultural sector. In recent years, however, several of these crops (the main ones being rubber, coconut and oil palm) have been providing 'forest' products that have been used for wood and fibre. For example, rubber wood is now used for the manufacture of about 80% of the furniture made in Malaysia, while coconut and oil palm trunks and the branches of rubber wood are used for various forms of reconstituted 'wood'. Rubber wood and coconut stems are derived from the conversion of old plantations formerly disposed of by burning, while oil palm fruit residues are used for medium-density fibre board. The development of these new markets has thus not only improved financial returns but also used the resources in an environmentally friendly manner.

The area of these species appears to have increased from the 14 million ha reported in 1990. The increase may be due to better coverage of the data, although it is known that oil palm areas are increasing rapidly, rubber tree areas are also increasing and coconut plantations are decreasing. Coconut plantations comprise the largest area (about 42% of the total), rubber 36% and oil palm 22%. Most of the coconut plantations are in Indonesia (33% of the area) and the Philippines (28%); most of the rubber plantations are in Indonesia (34%), Thailand (20%) and Malaysia (18%), while most of the oil palm plantations are in Malaysia (44%) and India (29%). Not all of the areas mentioned above are suitable or available for substitute 'timber' production but they illustrate the potential.

Contributions of Forest Plantations to Wood Supply

The continuing increase in the area of forest plantations which have been established for industrial wood supply has been to meet the reduction of outturn foreseen from natural forests arising from deforestation and changes in land use (largely in the tropics and subtropics) or from natural forest being taken out of production and devoted to service functions such as conservation. It has been believed that the outputs from forest plantations can help to reduce the pressure on natural forests as sources of industrial wood supply; while the logging of tropical natural forests is not the prime cause of deforestation, logging roads often provide the means for farmers to gain access to forests. The reduction of logging, combined with effective protection, may thus help to reduce deforestation in certain locations until land use and ownership are clarified. However, none of these palliatives will remove the underlying causes of deforestation: high rates of population growth, poverty, hunger and a shortage of fertile land to cultivate.

The potential of forest plantations to meet demand for industrial roundwood is considerable; it has been estimated that the present global demand for paper pulp could be met from an area equivalent to only 1.5% of the world's closed forest area. No global estimates of current output of timber from forest plantations are available, although FAO's global fibre supply model (GFSM) estimated that the potential annual growth of industrial wood from forest plantations in developing countries was about 5% of the increment of natural forests in 1995. In some countries, plantation production already makes a highly significant contribution to the industrial wood supply, for example in New Zealand 99% of industrial roundwood in 1997 was grown in plantations, while in Chile the equivalent figure was 95%, in Brazil and Argentina 60%, and in Zambia and Zimbabwe 50%.

Estimating the future contribution of forest plantations to wood supply is at present imprecise and is based on many more or less unreliable assumptions, particularly concerning the rate at which afforestation will continue. By the year

2010 the GFSM (op. cit.) estimated that the potential increment from forest plantations would be about 40% of that from natural forests in Asia, Oceania and Latin America and about 15% in Africa, under rates of deforestation and afforestation largely the same as today.

FOREST LAND SURFACE

The world's forests cover an area of 3454 million ha or approximately 26.6% of the land surface, with 56.8% of this area in developing countries, which are mostly tropical. Forest cover has largely stabilized in most industrialized countries but deforestation continues in many developing countries. Between 1990 and 1995, the area of natural forests in developing countries decreased by an estimated 13.7 million ha per year, although this rate of loss appears to be slightly less than in the period 1980-90. Furthermore, forest management is relatively little practised at present in natural tropical forests. The extent and condition of the global forest resource are determined by many economic, social and political factors external to the forestry sector, including, in particular, continued population growth and higher rates of global economic growth. Population growth, which will occur mainly in tropical developing countries, will continue to be combined with urbanization, while global economic growth will continue to be combined with changing consumption patterns, especially in the regions with fastest economic growth.

In the coming decades, pressures for increased food production are expected to lead to continued conversion of forest land to agriculture in many developing countries. It is estimated that in developing countries 90 million ha of land, of which about half may be forest land, would need to be converted to arable crop production alone between 1990 and 2010. Agricultural land expansion is projected to be faster in sub-Saharan Africa than in the past; given the unsuit-ability of much of this zone for agriculture, there must be a continuing threat to natural forest cover from agricultural expansion as land continually goes out of agricultural production. Infrastructure development will also contribute significantly to the continued loss of forests. In addition to deforestation in developing countries, large areas of forest worldwide are being degraded by overharvesting, overgrazing, pests, disease, wildfires and airborne pollution.

At the same time as forest cover globally is decreasing and forests are being degraded, demands on forests to supply wood and non-wood products and social and environmental services are increasing. Global consumption of wood increased by 36% between 1970 and 1994, and is expected to increase by another 20% by 2010.

Greater emphasis is being put on the services that forests and trees can provide, including soil and water conservation, sequestration of carbon for the mitigation of climate change, conservation of biological diversity, support in combating desertification, enhancement of agricultural production systems, improvement of living conditions in urban and peri-urban areas, and provision of educational and recreational opportunities.

Forests will remain an essential source of the livelihood of the poorer sectors of the world's population and a home to indigenous peoples for some time to come. The challenge of meeting the growing demand for forest products while safeguarding the ability of forests to provide a wide range of environmental services will increasingly be met through the planting of trees, either within the forest or outside. However, less natural or seminatural forest will be converted to plantations due to the emerging values of such ecosystems, and suitable land for plantation development will therefore be in short supply in many places. Governments will thus aim to encourage large- and small-scale landowners to plant trees outside forests, often integrated into agricultural systems, through policy measures including incentives. There are likely to be moves towards making blocks of trees more 'natural' with a diversity of species, in some instances in order to provide a wider range of goods and thus serve as an insurance against the possibility of a single species failure or to guard against possible loss of soil fertility or site degradation, as well as to provide for improved amenity and recreational potential. There should be less emphasis on the area or quantity of forest cover and more attention paid to forest health and condition.

Forest cover in industrialized countries will continue to expand, and several newly industrialized countries will see their forest cover stabilize and even increase. Industrialized and newly industrialized countries will place greater emphasis on the conservation of natural forest where it still exists and on the conservation of seminatural forest. Plantation areas will continue to expand, either as intensively managed systems akin to farming practices or management will tend to move towards a more 'ecosystem' approach.

RESEARCH FOR PARTICIPATORY RESOURCE MANAGEMENT

Research for participatory resource management requires, but is not limited to, the use of participatory methods. In other words, PNRM does not mean that only participatory research approaches and methods can be used. A wide range of research methods, both participatory and non-participatory are combined and need to be understood as a spectrum of methods and approaches from which stakeholders – not just researchers – can choose. The cases analysed in this book illustrate how research for participatory management involves stakeholders in generating new information relevant to making decisions about the parameters and procedures for adaptive management. These parameters or procedures may include the boundaries of the ecosystem, the relevant actors, the physical and social spaces for intervention, the priority problems and opportunities, the alternative development paths, optional interventions (both technical and institutional) and the tradeoffs these entail for different stakeholders.

The Schreier case study in this volume illustrates the combination of geographic information systems (GIS) research with participatory management; Vaughan's case study in this volume shows how modelling is being integrated with participatory research methods; Martin and Sutherland explain how researchers' own institutional studies were used to inform participants in community meetings convened to vision new forms of devolving resource management. Snapp and Heong, in this volume point out that a major research challenge is to combine the various 'information bits' derived from different stakeholders, and distil these into decision rules that they can use.

A useful rule of thumb is: the more stakeholder 'buy-in' that is required – and the more diverse expertise needed to generate the information required to reach agreement – the more important it is to use participatory research approaches and methods for NRM.

However, whether participatory or non-participatory methods of enquiry are used, research carried out for PNRM has to incorporate stakeholders' different research objectives and criteria for validity and credibility even if, for example, stakeholders are not involved in data collection and analysis. Then methods – both participatory and non-participatory – need to be agreed upon that meet these objectives.

PARTICIPATORY RESEARCH

It can be seen from the above discussion that participatory research is a collection of approaches that enable participants to develop their own understanding of and control over the processes and events being investigated. This is derived from the principle that greater understanding and power to use information results from being involved in its generation. Participatory methods for monitoring and evaluation help to make NRM more accountable to stakeholders, and to give participants greater confidence in the results. In the McDougall case study in this volume, easily understood criteria and indicators are developed by local communities, researchers and other stakeholders. These provide a framework for later monitoring, and for assessing key factors and their direction of change. This monitoring process creates the opportunity to feedback information and learning into the community forest management system. It thus serves to guide future action, helping to increase the sustainability of community forest resources. In a different approach, Vernooy and McDougall show how creating a set of environmental monitoring indicators with stakeholder participation, and presenting these to local government decision-makers, raised awareness and provided a basis for action. Participatory *action* research has the added objective of enabling participants to act more effectively based on their own improved understanding. Action research combines intervening in the process being studied with investigating the changes this action produces, and this approach is highly compatible with the concept of adaptive PNRM.

Different kinds of participation in research are possible and there are several typologies that distinguish along one or more dimensions (Arnstein, 1969; Biggs, 1989; Pretty, 1995). Empirical study of how different kinds of participation are being used in participatory research shows a huge diversity of practice in combining different types. Analysis of 150 NRM projects using participatory research shows that there is a definite pattern of using more empowering types of participation in the dissemination of results – ie, at a stage when conventional researchers are most comfortable in 'letting go' and relaxing conventional controls.

One of the fundamental differences among approaches to stakeholder participation in research for NRM is the way in which power relations among different stakeholder groups are structured. Natural resource management and research about it are embedded in power relations. These may encompass powerful international, national or regional interest groups and relatively powerless local people. And they may also include the relatively wealthy, high caste or male members of a community in contrast to the poor and those of low social status, such as women and minority ethnic groups. Information is one source of power in a changing NRM situation, and participatory research can purposively generate new information that changes the balance of power, and can strengthen the bargaining situation of less powerful stakeholders.

Processes to promote participation in the management of resources and in research that fail to examine how power relations affect, and are affected by, the participatory process are often superficial and transitory. For example, participatory rural appraisal (PRA) has been heavily criticized for failing to recognize the incentives for different interest groups to manipulate the appraisal process. The way in which power relations among stakeholders are handled in a participatory research process is intimately related to the issue of research quality. For example, gender relations affect the distribution of power in a participatory research process and bias results.

One way to assess quality in participatory research is to ask: 'How valid and reliable do the different stakeholders who are party to the research process judge the results to be?' In a PNRM process researchers are stakeholders who set research standards, but they are not the only ones. Thus standards for reliability and validity have to be negotiated with stakeholders. Often researchers have to accept compromises. A variety of different standards will often have to be met for quality assurance. The way in which power relations shape results can make or break the credibility of both the research process and its conclusions. For example, Mosse (2001) describes a PRA sponsored by a State Forest Department in India in which an overwhelming preference for planting eucalyptus trees was identified among the villagers participating. It turned out that villagers had little knowledge or experience of eucalyptus, but prioritized what they perceived the agency was able to deliver. The results of the PRA reflected the balance of power between the villagers and the State Forest Department, but did not provide a valid assessment of villagers' needs nor a conclusion that stands up to further analysis. Sutherland's case study in this volume reports that villagers did not distribute vetevier grass planting material from experiments to other communities because they did not have permission from the project that had paid them for growing these, illustrating how power relations also affect ownership and how research results are used.

One of the major threats to the validity of research occurs when stakeholders have not explicitly negotiated how control or ownership of a participatory research process is going to be managed. Ravnborg et al (1996) show how the exclusion of a key stakeholder group from a problem diagnosis led to a result that was fundamentally biased against them and towards an interpretation that ultimately damaged the agreed-upon collective reforestation programme. Only once a forum was created, in which the absent stakeholders were included, were new information and competing interpretations of the advantages and disadvantages of slash and burn practices aired. Only then was it possible to negotiate a viable plan for collective action, which was subsequently successfully implemented. The cases in this book illustrate the broad spectrum of approaches to managing power relations, control and ownership of participatory resource management and the research it involves. For example, the African Highlands Initiative (AHI) case did not negotiate power relations explicitly and this affected research quality as researchers began to drop out because of their loss of control.

The 'Landcare' process is quintessentially owned and driven by local groups, but managing the dynamics of power relations among different stakeholders within the Landcare groups or among groups, and how these affect research is not evident in the project strategy. In the Centre for

International Forestry Research (CIFOR) case study reported by McDougall (this volume) researchers were flexible from the start in providing a framework for monitoring that enabled multiple stakeholders to develop their own indicators, and eventually to take over and adapt the framework. This approach focused on generating feedback and adaptive learning, making it easier for researchers to see the advantages of 'letting go', but the negotiation of power relationships among other stakeholders in the research process was less explicit. In all these cases, the motivation for the participatory research is researcher-driven, at least at inception.

A different approach is illustrated by the explicit negotiation of control in a PNRM process, which is secondary to and embedded in solving a compelling wetland management problem. This case initiated its process with a meeting in which a vision for the future of the wetland involved various 'stakeholders' presenting their individual views and negotiating a basic agreement on rules to be respected and activities to be carried out. Within this framework of agreed rules, stakeholders decided what research they needed and how to collect it in order to throw light on different resource management options.

In summary, research for NRM cannot be carried out as if it were independent of power relations among researchers, or between researchers and other stakeholders. For this reason, a capacity for organizational learning in research organizations is an important determinant of the outcomes and impacts of research, because organizational learning is essential for transforming power relations that otherwise become an obstacle to innovation in NRM, as the case study by Stroud in this volume illustrates.

The models or theories of participation and resource management that drive innovation in a research organization engaged in NRM are critical determinants of research practice. Many of the problems encountered in conducting participatory research are rooted in organizational behaviour rather than in the choice of methods or types of participation.

Common Organizational Problems in Participatory Research

- Lack of representation of key stakeholders in the research process.
- Participation is not developed around clearly specified rights, roles and responsibilities.
- Mechanisms of accountability among participants are lacking, especially the accountability of researchers.
- The process is corrupted by hidden agendas.
- Conflicts of interest are not made explicit nor negotiated.
- Transaction costs of participation exceed the benefits to the participants.
- Feedback mechanisms, such as monitoring and evaluation of the research process are not in place so that learning about how to improve the process is minimal or slow.

Models of participation in a research organization provide a means to structure and organize the research process, methods for decision-making, and the rules and behaviours of researchers. Models that are incompatible with adaptive PNRM make it difficult for the necessary learning to occur. When models of participation incompatible with adaptive PNRM prevail in a research organization, for example when research and development is the dominant model in contrast to research *for* development, most of the innovative research is done by an informal, or 'shadow', organization that develops as a way of circumventing the outmoded rules of the formal organization (Sherman and Schultz, 1998). Rocheleau, in this volume, analyses the world of isolated and undocumented participatory research outside formal research organizations and the institutional divides within formal research that are an obstacle to organizational learning.

In contrast, in a learning organization new models of how to conduct research *for* development are rapidly incorporated and innovations are readily undertaken. The idea of a learning organization arose in the private sector out of the need to be adaptive in the face of rapid change driven by intense competition, and the learning organization concept has several features in common with participatory methodology. The models of participation that drive organizational behaviour and research practice are based on underlying principles (defined as the ideas that are used to formulate models). Principles are more important than rules or methodology: 'rule-generated behaviour doesn't work' (Sherman and Schultz, 1998). For this reason, an important focus of this book is the illustration of the underlying principles of participation that are more important than the specificities of one or another participatory methodology. One way of illustrating how principles are more important than methods is to examine how participation in research adds value to adaptive NRM.

10

The Wealth of Forest Tree Species

WOODY TREE SPECIES

The earliest forests of the Carboniferous Age were probably not composed of woody tree species but of ferns, cycads and other vascular plants that were neither conifer nor angiosperm. The more complex vascular and reproductive systems of coniferous species did not evolve until the Devonian Period some 350 million years ago. Conifer trees were not prominent until the Permian Period 100 million years later. It was another 70-100 million years before the pines differentiated, around 180 million years ago. The soft and hard pines diverged around 130 million years ago, during and following the break-up of Pangaea and the great continental migrations. Angiosperms evidently did not originate until the Cretaceous Period, but soon broadleaf tree species would become more prominent in forests. By around 100 million years ago they began to dominate over the ferns and gymnosperms in many forests.

The end of the Cretaceous Period, around 70 million years ago, is marked by major animal extinctions, wide climatic fluctuations, continental shifts with episodes of mountain building, and the differentiation of the major angiosperm family groups and genera. Periods of major extinctions also occurred during the Permian and during the global cooling that took place at the end of the Tertiary. Further speciation continues in both gymnosperms and angiosperms to this day, but largely on the development plans established by the end of the Cretaceous.

Since most of the major families and genera existed at least as prototypes by the end of the Cretaceous Period, there are large patterns of geographical differentiation in species, genera and families. Among the nearly 600 species of conifers, the Pinaceae (10 genera, 250 species) is a Northern Hemisphere family, as is Taxaceae (5 genera, 15 species), Cephalotaxaceae (1 genus, 5 species) and Taxodiaceae (9 genera, 12 species, with one southern genus with 3 species), while Podocarpaceae (7 genera, 131 species) and Araucareaceae (2 genera, 36 species) are predominantly Southern Hemisphere families.

Cupressaceae (16 genera, 142 species) is spread over both hemispheres. All species are monoecious and wind pollinated.

Among the angiosperms, a much larger number of species (20000-25000) evolved during the Cretaceous and the development of the progenitors of many presently known species continued through the early Tertiary Period of large climate change and widespread extinction. A wide variety of reproductive behaviour evolved, including monoecy and dioecy, and all forms of sexual divergence in between (Bawa & Opler 1975; Charlesworth & Charlesworth 1978; Charlesworth 1984, 1989), which affected the rates of population and species divergence and evolution. However, the establishment of mating barriers is not the *sine qua non* of speciation for plants as it is for animals, and other forces have relatively more effect on the creation of new species (Grant 1971; Grant & Mitton 1979). To understand how the present array of tree species arose and how future diversity may evolve, several evolutionary factors must be considered.

The selective forces of the environment may favour different inherited traits and force divergence. Mutations may also generate diversity, and the reproductive system itself may erect barriers to intermating among the derived and the progenitor groups. Other events such as geographical isolation may form barriers, and further isolation could reinforce the effects of selection and mutation to generate distinct species. If further divergence occurs, then species-level differences can accumulate to generic and familial levels of divergence as later speciation events accumulate differences and sets of descendants diverge further. Such divergence may be rapid or slow depending on the strengths of the evolutionary factors, but all factors were probably strong simultaneously during those periods of rapid speciation such as occurred at the Cretaceous-Tertiary boundary.

DEVELOPMENT OF SPECIES

The development of species (cladogenesis) is seen as a process that proceeds from the normal evolution of a species which, if no split occurs, would be called anagenesis, or the evolution of a species lineage without subdivision. Within a human lifetime and on a local scale of observation, microevolutionary changes composed of shifts in species traits can be observed, such as when tree breeding programmes change growth rates or stem form. Species seem to remain fairly constant in their general form and behaviour, though with gradual shifts in a continuing evolution of adaptation (anagenesis). Evolution may be largely dominated by events common to the whole species, if populations are often intermating and subject to similar environmental conditions. Even if subject to different and strongly selective environmental conditions, occasional interpopu-lation mating can force their evolutionary dynamics to be very similar to that of a single population. However, on a longer time-scale than included in a single human generation, or on a large geographical scale, species may not always evolve as a unit with gradual continuous steps.

On a global scale, it is expected that environmental differences are so large that the segregation of different adaptive capacities would often be advantageous and hence that selection on heritable traits would be a factor in speciation. Since ecologically critical factors are not distributed uniformly around the world, divergence rates among species would also not be expected to be uniform. We can see some of the broad strokes of large effects in the contrasts between the boreal and tropical forests. The obvious factors of temperature and moisture are so different among broad regions that divergent adaptations, at least to major climate extremes, would clearly be advantageous. In addition, it can also be observed that in the boreal forests most species carry both sexes on the same tree, have their pollen carried by wind and are usually found in relatively large continuous stands.

In contrast, tropical forest trees often carry males and females on different trees. These species are most often sparsely distributed throughout most of their range (Ashton 1976; O'Malley & Bawa 1987), and insects, birds or mammals carry pollen. Opportunities for inbreeding or for restricted localization of mating neighbourhoods are much greater in such species. Tree species differ in their reproductive behaviour as well as in the environments to which they are better adapted. It can therefore be expected that forest regeneration dynamics would differ among regions and that their genetic dynamics would also differ, leading to segregation into divergent populations or subspecies.

Among the angiosperms, many reproductive systems exist, from those that are primarily self-pollinating to those which are primarily outcross-ing. Pollination may be predominantly mediated by wind, insects, birds or mammals, and seed dispersal may be proximal or distant; both may be episodic between years or may be continuous throughout a year. The mating system of some species is conducive to restricting the pool of potential parents. The mating system may also force some species to depend on one or other type of pollinator and would lead to reproductive isolation. In addition, the possibility of the mating pool being very limited would, if coupled with a non-continuous ecological distribution, allow some species sets to have very rapid rates of evolutionary divergence.

Thus if ecological conditions act as selective agents that differentially affect survival or reproduction of some genotypes, the genetic composition of the population may be biased by selection for tolerance to competition or by sexual selection on rare genotypes. In such cases population evolution would be dependent on the combination of genetic and ecological events. The evolution of populations can then affect not only the immediate survival of that population but can also determine the future subspecies structures that may lead to further speciation or other genetic structure for the species as a whole.

Thus, the mating system, environmental differences and history can all affect the rates of species creation by cladogenesis. Cladogenesis and anagenesis would be expected to often follow the same rules of evolutionary dynamics whether in single or multiple populations, and to be determined by the effects of mutation, migration, selection and population size.

The forces of evolution act on the capacity of a species to respond to challenges of the reproductive or vegetative environment, and that capacity is determined by its genetic endowment. This genetic endowment determines what it is that we call a species at any one moment in its evolution and also determines what its evolutionary trajectory may be. Therefore, discussing the species present in forests requires discussion of what their genetic endowment is, as well as what forces are acting on it. Discussion on how species are changing, or can be conserved, also requires discussion on how their genetics is changing and how their genes can be conserved. Since many of the dynamical events that affect evolution occur among individual trees, the individual is considered to be the primary unit of evolution, and the genotype (the collection of all the genes in an individual) is the unit of gene expression. The phenotype is the physical expression of the genotype within the environment of the individual's lifetime and is the unit of selection. The correspondence between individual-level events that we can observe and the long-term evolution of species groups, which takes a longer time, is the realm of evolutionary and population genetics.

However, evolution is not a unidirectional phenomenon, with trees and their genes passively exposed to a selective environment. It does not proceed solely from a given set of environments that determine which genes are present, nor solely from genes to how the trees will perform. At a large level, it is clear that the forces which affect reproductive behaviour and selection are themselves changing partly as a result of the genes that are favoured. As climates and human influences change, the ranges and density distributions of species can also be expected to change. The migration of species may thus be driven by general environmental influences, but would also be influenced by how other elements of the biota abet or hinder their movements. These are factors that are partially dependent on which genes are present. For example, in the temperate and boreal forests the postglacial movement of species would not be expected to be uniform since they differ in colonizing ability and their initial distributions.

In addition, some species would form migrational barriers to other species and hence the history of migration would also affect the distribution of species and would inhibit the existence of an optimal composition of species in communities. One pioneer species will often exclude other potential pioneer species, and secondary succession may depend on those initial events as well. Thus, tree communities may in fact be associations of species that have not coevolved to their mutual benefit but coexist due to independent adaptations to similar environmental conditions, or that jointly occur for ephemeral and historically dependent reasons.

Evolution within forest tree species may therefore be largely independent of the evolution of other tree species. They are less likely to be independent of interacting species such as their pollinators.

The dynamics of forest ecology depends on the characteristics of the species that can invade and maintain themselves, and is therefore partly a result of the species endowment available. However, the evolution within species also depends on those same ecological factors. While it may be obvious that major differences in forest dynamics exist among the forest zones, all forests go through periods when some species establish themselves in new populations while others die out, at least in local areas. On a given area, the displacement of one species by another is a common occurrence that can be triggered by extrinsic forces such as fire or by autonomous factors such as competitive exclusion of one species by another. Some of these displacements are mediated by differences in species adaptations to regeneration under shade vs. open areas or to differences in competitive ability for soil, water or light resources. Some displacements may be due only to the accidents of the timing of seed arrival or other random events.

While a forest cover may exist in some areas for many centuries, at least local variations in species composition over time would be the general rule and hence the local regeneration of any one species iswould be expected to be often sporadic and episodic (Loveless & Hamrick 1984; Shugart 1984). Even for forests composed of shade-tolerant and self-replacing species, gaps frequently disrupt the continuity of forest cover and generate sequences of displacement.

Some species, called pioneers, are also dependent on large gaps or cleared openings for regeneration, and normally form dense colonies in the foundation of populations. The seeds of such species may be easily transported over long distances or may be stored in a soil seed bank, enabling cleared areas to regenerate a forest cover. Their mode of regeneration also implies that they suffer local extinctions, through their own effects on their environment, through displacement by superior competitors or by environmental catastrophes. Forest dynamics itself brings evolutionary pressure on the forest's constituent species.

Selection Advantage

The selective advantage of one individual over another is one of the main forces of evolutionary genetics, and along with mutation, migration and drift form the genetic endowment of a species and of the populations of which it is composed. Evolution from the individual level of selective and reproductive events is one of the ways that we can understand the transition from what is called microevolution to higher taxonomic levels, or macroevolution.

Competition dynamics is often invoked as one of the strong selective factors in forest tree evolution. Species that benefit from or tolerate shade as seedlings or saplings may occur in higher frequency under a canopy, or may invade areas already occupied by other vegetation. They may emerge into the upper canopy when openings occur and hence the distribution of reproductively mature individuals would follow patterns established by the size and frequency of canopy openings. Some species in tropical forests require shade or a protective canopy cover and can reproduce only under a canopy where their crowns can form a low canopy layer. However, light is not the only environmental limiting factor that affects the distribution of trees. Other factors such as soil type and soil moisture can have a strong influence on the extent and patchiness of a species distribution.

One of the basic necessities for evolution is the capacity to adapt to new stresses in the physical or biotic environment. For any selection to lead to adaptive evolution, there must be genetic variation in the stress response and the trees with the favoured trait expression must leave more progeny.

Since many environmental factors are themselves changing in response to extrinsic forces, we can expect that stand composition and species distributions are perpetually changing for extrinsic reasons. Thus, selection factors are not expected to be consistent over time. In addition, historic influences on the recruitment pool for species establishment can affect the probabilities of species replacements and simple stochastic variations on the demographic distribution can also influence distribution for reasons not related to optimality of fitness. Hence, the accidents of history, stochastic variations in environmental variables, lag times in selective effects, as well as deterministic causes affect the actual distribution of species.

Virtually all species can be considered to be involved in a dynamical system, to still be migrating both geographically and ecologically, and to be evolving in response to directional evolutionary forces as well as to both deterministic and stochastic forces. Therefore, forest tree species can be expected never to be in an optimally adapted state or, if so, only transiently.

Thus, while selection can be a significant force in the evolution of species and in the evolution of diversity among species, it has effects within a complex ecosystem and on complex organisms, and its effects have a lag time. Its effects also depend on the genetic endowment and structure of variation already present in the species and on the other forces of evolution, including mutation, migration and drift. These forces can change the species characteristics and if sufficient genetic variation is maintained within or between populations, continued evolution can maintain the species and can sustain forests. *Sustainable forestry and the maintenance of species thus require a sustainable genetic resource in some form.* The wealth of genetic variation within and between populations and the combined genetic and ecological dynamics thus jointly determine the inherent structure of forests.

SPECIES MIGRATE AND CHANGE THEIR GEOGRAPHICAL DISTRIBUTIONS

If we examine how species migrate and change their geographical distributions over a few generations, the establishment and extinction of local populations is not an unusual event. If such populations become reproductively isolated and if, either by chance or through selective events, they also genetically diverge in mating behaviour and adaptability, the basis for developing a new species exists. If populations do diverge, they would be likely to do so at many genetic loci and at many alleles at the same time. Hence, whole suits of traits and sets of genes would be expected to shift together as populations segregated into different mating groups. Not all of these changes would be ascribable to selective advantage of the alleles, but rather to chance events and to accidental associations of genes and traits.

In the relatively simple boreal ecosystems, the movements of populations escaping from glacial encroachment into sufficiently hospitable areas and alternately advancing into retreating glacial boundaries would be expected to allow populations to carry adaptations to conditions that lag behind actual environments. Nevertheless, during episodes of environmental change or of species movement into new environments, some traits would be of adaptive significance and near optimality, while others may be of little significance or may lag far from any current optimality. Only if the populations carry sufficient genetic variation could they respond rapidly to optimizing selection and ultimately achieve or approach an optimum condition, and then only if the environment is also stabilized.

During periods of change, populations may diverge from a base population or from each other in location or habitat. For divergence to become inherent among them, strong selection on populations with genetic variation in traits of adaptive significance and an almost absolute barrier to pollen or seed migration between populations would be required. For *Pinus con-torta,* strong interpopulation differences can be observed despite substantial gene migration. Most often, it can be expected that without designed or accidental interventions, separated populations would sooner or later reconverge and that periodic intermating among populations would be frequent enough that traces of different gene frequencies and linkages would be difficult to detect. If populations once diverge, they may often simply integrate upon secondary contact and re-form as a single intermating population. Without strong mating barriers for a long time, the long-term evolution of the species as a whole would be a slow approach to an average optimality by the increments of mutations to new alleles at some loci.

Occasionally however, divergence is strongly established for long enough that at least partial mating barriers are formed. Then, splitting of a single former species into two or more branches could generate new species. While the occasional speciation event may not be often repeated, incipient speciation could generate stronger and stronger divergences in genetic composition as selective differences and unique mutations accumulated. Once a coherent species is formed, any further speciation events would follow on the basis of prior divergence and there could then ensue a hierarchical series of branching species that ultimately develop into genera and family level divergence. This is the basis on which phylo-genetic relationships are constructed and which underlies the power of molecular genetic data for reconstructing evolutionary trees.

The existence of large populations undergoing population isolation and of changing environments were pronounced during the Cretaceous Period, when many of the major forest tree speciation events occurred. Subsequently, this led to the generic and family level structures that we see today. Though the potential for speciation events continues to exist through modern times, the events occur at slower rates than humans are attuned to observe.

However, by using experimental evolutionary approaches, including strong selection and the construction of strong mating barriers, such as among divergently selected breeding populations, it is possible to create large population divergence both genotypically and phenotypically. As long as new species arise by splitting parental species, and each accumulates random mutational changes, divergence among species would follow patterns of statistical distributions, and by tracing backwards in time, similarities can be used to reconstruct coalescent histories.

OCCASIONALLY SPECIATION EVENTS IN PLANTS

Chromosomal Change

Occasionally speciation events in plants also occur as a result of reorganization of the genetic material, especially by means of chromosome rearrangements and duplications.

For example, it is reasonable to conjecture that chromosome rearrangements explain the number of chromosomes of Douglas fir, which has 26 chromosomes in its vegetative cells instead of the usual 24 carried in all other conifers. This is probably the result of one of the usual conifer chromosomes dividing in two and developing its own centromere. In tropical hardwood species, it is not uncommon to find irregular series of chromosome numbers, seen in the dipterocarp species especially.

Within most families, the basic haploid number of chromosomes (x) is the same for all genera and species, but for some, as among the Aceraceae and Betulaceae, there is a wide variation in chromosome number among genera and species (Grant 1971; Morgenstern 1996). There is a series of species with chromosome numbers that are multiples of the basic number (*e.g.* $2x$ to as much as $6x$) which form distinctive species. In aspens, occasional triploids arise following species hybridizations. The genes in these species may simply be replicated among the chromosome sets and all continue to add to gene expression.

However, the kinds of interactions among alleles can be affected by their pairings as well as by the numbers of alleles at the same 'locus' and among 'loci'. Such a series of chromosome multiplication is a polyploid series, and may arise as the result of a meiotic failure of the original nuclear genome to split into two separate nuclei. This may occur often enough within a population that the pollen or egg cell can mate with another that also carries the doubled set, and instead of producing a diploid would produce a tetraploid. If a diploid germ cell is paired with a normal haploid cell, a triploid would form, although these usually fail to generate viable germ cells. However, if a diploid and a tetraploid happen to mate and those cells double their chromosome set, a hexaploid could result.

The creation of polyploid series may be stimulated by hybridization events, where chromosome mechanics are sufficiently disturbed that unusual meiosis events may occur more frequently. However, hybridization events by themselves can also create new kinds of species dynamics such that new species are generated even if the chromosome numbers are the same and do not otherwise change.

Hybridization

Hybridizations among divergent species occur when they are not too divergent for the intermat-ing to occur but are distinctly different in form or function. They may be frequent enough among plant species that their potential for shaping evolutionary relationships cannot be ignored. Hybridization occurs with much greater frequency among plant species than among animals. Examples of hybridization in fruit trees can be found with different outcomes. At some point during the divergent development of distinct species, a barrier may exist that further reinforces intermating such that no hybridizations may occur, although this seems to be at fairly wide levels of divergence. Before that time, hybridizations can intermingle the genetic endowments of the incipient species and can thereafter serve as foundation taxa for further evolution of divergence among species. Therefore, the existence of hybrids and populations of intermediate genotypes implies that the speciation process has not gone so far that new species can evolve by hybridization. In fact the processes of divergence and hybridization are continuing to this day. The large areas over which lodge-pole pine and jack pine hybridize indicate a broad compatibility of distinct species that may stabilize in some areas, but with intermediate types developing as a new taxon.

In some cases, the population swarm that characterizes recent hybridizations may segregate the parental types as second-generation intercrosses or as backcrosses to their parental populations. Alternatively, they may maintain intermediate types in most traits and eventually stabilize as a taxonomic unit with some intermediate trait expression and with some mixture of different parental traits.

In the evolution of a hybrid population between *Pinus taeda* and *Pinus palustris* created by the natural intercrossing of the parental species in the 1920s, later generations failed to maintain any intermediate types and, 20 and 40 years later, when their progeny reproduced they predominantly gave rise to the parental types. Apparently, the original hybrids were successful only in the peculiar environment of abandoned farms from which fire was excluded; when a forest cover was established, with occasional brush fires, the hybrid either was not selectively stable or was chromosomally unstable, or both. In contrast, the hybrids between *Picea sitchensis* and *Picea glauca* form a large and apparently stable zone of introgression in northern British Columbia and hence can be stabilized as a hybrid with backcrossing to both parental species. The hybrid types may disappear or may form a permanent zone for hybridization, or may eventually form a third 'species'.

If just two parental species are involved in the development of a hybrid, it might be difficult to distinguish if an intermediate type is a recent hybrid of historic species or if the intermediate is a historic source population for newly divergent taxonomic units. If historical evidence is available on prior migrations, then the origins of the units can be readily distinguished by the times of establishment. However, in the absence of historical data the intermediate may be either a hybrid progeny of two divergent parents or the parental source of divergent progeny. If genetic measures can be obtained for multiple, independent and neutral loci, it would be expected that the isolation of divergent populations would initially involve a random segregation of genes from the original population.

This is a case of single source population producing two derived branches, and instead of simply splitting into two, three branches may exist from a single origin. Then the derived progeny populations would receive an admixture of alleles such that among the progeny populations themselves there would be similarities in some traits and differences in others. If alleles or traits were different between the progeny populations, they would have diverged from the parent population but would not necessarily diverge in symmetrically opposing directions. The allelic richness of the source population would be expected to be at least as great as that of both of the derived progeny populations, each of which would contain subsets of the parent population's alleles.

In contrast, a recently derived hybrid population would share most alleles at frequencies between the parental populations and would therefore be intermediate in most traits simultaneously. The divergent populations then would seem to diverge in symmetrically opposing directions from the intermediate population at all divergent loci and in all traits. The hybrid population would also be expected to have alleles that exist in at least one of the parent populations but not to contain all of the alleles of both parents. The eventual divergence of hybrid populations can, of course, involve backcrossing and selection such that they ultimately cannot be distinguished from the previous model of divergence from a parental stock. However, linkage disequilibrium in the hybrid population should linger and be detectable for many generations. In this case, two parental species can give rise to three species if the hybrid continues to develop and establishes mating barriers such that it can evolve independently. In the case of *Pinus rigida, Pinus taeda* and *Pinus clausa,* there are three two-way hybrids between the species that could indicate a close relationship between them and a splitting of an original one into two or three by a subsequent split. However, it could also imply a recent hybridization of at least two of the parental species.

These forms of speciation can lead to further diversification of lineages or to convergence of genetic lineages, if some sets of genes are transferred among several species. These forms of speciation may then form multibrachial taxonomic structures and cross-linkages of sets of genes, neither of which conforms to evolutionary trees. The existence of three-way hybrids in pines, for example, indicates that such complexities in speciation may not be unusual at least among the pines and may not have been unusual in the development of the family. The construction of genetic lineages therefore is not identical with the construction of species lineages, and local events in the evolution of populations can have significant effects on long-term phylogenies.

The events involved in speciation do not generally occur in all units of a species at the same time, but instead involve genetic events that change individuals or that occur among population isolates. If species-level events are contained within a matrix composed of other species, they may often exist as a set of reproductively isolated islands, and if the parents of new populations rarely mate with members of other islands, the species may exist in a metapopulation structure of loosely connected subpopulations. This may even lead to incipient speciation as the islands become more distinct if they further form mating barriers among populations. Thus, a species may be distributed in the form of large, indistinct, continuously intermating populations, or as discrete and distinct metapopulations and in mixtures of large and small populations with partial migration among some subsets. The amount and distribution of genetic variation will be influenced by the metapopulation structure as populations split into large or small segments and as they either advance or retreat from habitable sites.

PLANT SPECIES, GENETIC VARIATION COMMONLY EXISTS

THE WEALTH OF POPULATIONS

Within plant species, genetic variation commonly exists at a high proportion of loci and tree species commonly exhibit the highest levels of all plants. For the species surveyed, which are mostly boreal and temperate pioneer species, the levels of genetic variation within populations often seem to exceed those between populations. The lack of strong differentiation may be due to the wide dispersal abilities of pollen and seed, which maintain a degree of connectedness between mating neighbourhoods among trees over long distances. A few studies of tropical tree species that are not pioneers indicate considerable variation among populations of trees, which may reflect a finer subdivision of mating neighbourhoods.

Another measure of genetic variation is the heritable variation that exists in traits controlled by many gene effects simultaneously but in which individual gene loci may have small effects.

These are the so-called quantitative traits that can involve many genetic loci which act indirectly on the observed trait and hence have small direct effect. Most of these loci would not change much in frequency from one generation to the next.

Cumulatively, however, they could cause large changes in the trait expression. Thus, many of the traits selected for in breeding populations can shift mean expression by as much as one standard deviation in one generation even if there is little shift in gene frequencies at loci sampled by molecular means. Furthermore, since differences among populations in the frequency of molecular markers are more difficult to detect than differences measured in quantitative traits, traits of adaptive significance may be more obviously segregated among populations than molecular genetic measures may indicate.

Some traits and the gene loci affecting them may be under strong selection pressure that is divergent between populations, while other traits may be under mild or virtually no selection pressure. Some traits may be indirectly related to strong selection events, such as caused by fire, drought, or forest practices that leave few resistant individuals, but only weakly related to resistance mechanisms.

Other genes may not be presently involved in any recent selection events. Thus the genes associated with selection response may be under strong or weak selection according to the strength of the selection event and the degree of relatedness of the trait to the selection agent. There may also be multiple genes that affect a trait expression, and the combined effects of the genes may not be simply assigned to any simple action of any one of them. This kind of multigene interaction is called epistasis and it may cause multiple gene locus differences to be more pronounced at the level of trait observations than in single-locus allele frequency differences. These kinds of interactions may mask any obvious trends in gene frequency changes associated with selection trends.

Since the phenotypic traits themselves can be expected to be under a wide array of selection intensities and the genes associated with them can also have a wide range of influence on those traits, we can expect that there would be a wide range of selection pressures among the genes of any one population.

Thus, if the relative divergence of populations was tracked by weakly selected traits or loci, population similarities would predominantly reflect migration effects. In the absence of selection, even very low migration rates would tend to maintain similar allele frequencies among populations. However, if divergence was tracked using traits or loci that respond strongly to divergent habitat selection, then similarities would tend to reflect habitat differences. In any one population, strong directional selection would force disfavoured alleles towards the point of extinction unless maintained by an input from migration or mutation.

Among populations exposed to divergent directional selection, different alleles would be at high and low frequencies among them and hence their frequencies would be highly dissimilar.

If a large number of traits or loci are used and some loci reflect similarities where others reflect multiple patterns of divergence, then contrasting patterns of allelic distribution must coexist. Similarly, traits subject to different intensities of divergent selection would also display different degrees of divergence.

Clearly, if phylogenetic relationships were estimated using selectively significant loci along an environmental gradient, then genetic similarities or distances would be correlated with that environmental variable. Alternatively, if selectively neutral traits were used, the genetic distance measures would be correlated with migration distances, and the derived phylo-genies would be different. The existence of differing strengths of divergence has been observed in variations in the patterns of geographically related population traits in *Pinus armandii* (Ma 1989). Differences in the amount and geographical pattern of isozyme allele frequencies and in growth traits have also been found for *Pinus lam-bertiana* and red alder.

In addition to those forces of evolution, the mating system can have heterogeneous effects on the distribution of alleles. If pollen and seed have different migration distances or different direction of flow due to their migration season or the influence of different wind pollinator effect on the sexes, then the balancing effects of migration and selection would be different (Namkoong & Gregorius 1985). Even for selectively neutral alleles, differences between the migration paths of the sexes causes heterogeneity of allele distribution.

Genetic variation also exists at several levels of expression and the amounts and patterns of variation can depend on which measures of variation are used to estimate that differentiation. With currently available tools of molecular measurement, it is possible to measure very fine levels of difference so that individual trees (or humans for that matter) can be discriminated by unique changes in the fine structure of single genes. It is also possible to discriminate differences among local populations if they share maternal or paternal lineages, by using particular molecular genetic markers that distinguish among populations even though the populations are not obviously different phenotypically.

Furthermore, some genes are expected to be more influenced by the different forces of evolution and therefore the distribution of variation among genes may vary according to their function (Ma 1989). Therefore, there is no one preferred measure of genetic variation but several options, and the choice of which measure to use depends on the purpose of the measure and the type of gene being measured. For forest trees it is satisfying to note that the levels of genetic variation measured are usually among the highest in the plant kingdom, although their distribution and the differences among loci have not been well studied. If the frequency of heterogeneous associations is high, the evolution of forest species may be unusually complex.

Some forest trees have relatively high levels of genetic variation between populations in several traits and at several molecular loci, while others may have low general levels. Since the beginnings of interest in genetics, foresters have known about economically significant levels of heritable genetic variation among populations in traits relating to growth and soil and climatic adaptability.

One of the earliest tests to indicate the utility of population differences is recorded for *Cryptome-ria* in Japan in 1883. However, the existence of graded changes in traits is widely recognized in forest trees as clines and with high levels of variation among populations relative to genetic variation within populations. The variation patterns often differ depending on which traits are measured but growth rates and climatic adaptabilities are often considered to be optimized within local populations. While there is debate as to the generality of the concept that local populations are optimal, regular trends in performances are often observed.

The geographical patterns of variation among populations or provenances may vary from a single, large, contiguous and intermating population with gradations across mating and environmental neighbourhoods to one of discretely divergent populations with little relationship among subpopulations. Some species may be composed of a main, large, central population with satellites connected only to the central one and with only one-way germplasm flow, or may be a series of more or less equivalent-sized populations with equal and multidirectional flows. Some of them may exist along streamside corridors as single units with connections between drainage basins being more tenuous, while others may form island populations on the tops of separate mountain peaks or ranges.

Since the stability of the interpopulation structure depends on the stability of ecological dynamics, the level of population or provenance differences may fluctuate over time and geographical area. In turn, the response of populations to environmental divergence is contingent on the genetic dynamics that can occur within an intermating neighbourhood. For the most part, the potential for selective divergence among the populations is moderated by migration among populations. Only traits with strong divergent selection would maintain high levels of interpopulation difference.

THE WEALTH OF INTRAPOPULATION GENETIC VARIATION

Within populations, four forces (selection, migration, mutation and population size) determine the structure of genetic variation. The interaction of all four factors can be highly complex and their effects are determined by the mating system and the kinds of gene effects that exist. A problem in analysing even this level of genetic evolution is that the kinds of gene effects that exist depend on the effects that other forces of evolution have had on gene actions themselves. To consider how the forces can affect evolution, we can first start with each of them and then consider how their interactions can change predictions.

For selection to have an effect on speciation and the structure of forests, there must be a difference among genotypes in the traits affected by survival or reproduction events and the individuals must accordingly leave more or fewer progeny than other genotypes. The genotype is defined as the individuals with a distinctive ensemble of alleles at its gene loci that distinguishes it from other genotypes. This is a multiple gene locus definition, which is often simplified by considering every one of the loci to be independent in the occurrence of its alleles in both frequency and effect.

In this case, we can consider the genotype to be the summation of all independent loci and hence can examine gene effects and dynamics one locus at a time. This is, of course, an oversimplified view since genes are in fact linked on chromosomes and therefore have linkages that correlate their occurrences, and most genes affect processes that are combined in their net effect on traits; furthermore, most genes affect enzymatic pathways that ultimately affect several traits when measured at the level of whole organism performance.

These effects are called linkage, epistasis and pleiotropy, respectively, and tie the whole organism together into an integrated set of actions and functions that can survive and reproduce. While undoubtedly true in general, there is little evidence that individuals are so bound by such constrained genetic constitutions and physiological limitations that substantial independence among loci exists. The evidence for the existence of strongly selected, tightly linked 'coadapted gene complexes' is not strong and hence the model of independence can be usefully employed, even though it is surely wrong at least in detail.

Within a gene locus, many alleles may exist and the pairwise interactions among them are called the dominance relations. For a diploid organism, if both alleles at a locus are the same, the individual is said to be a homozygote; if the two alleles are different, the individual is said to be a heterozygote. If the heterozygote has a phenotype intermediate between the two homozygotes, the dominance effect is considered to be zero and the gene action is said to be additive. If the heterozygote is other than intermediate, it may be less than the lesser of the two homozygotes. If its phenotype lies between the homozygotes but is not exactly intermediate and if it appears to be greater than the greater of the homozygotes, the gene action is said to be underdominant, partially dominant or overdominant, respectively.

The effect of selection would depend on the kind of gene action, since directional selection for greater or lesser phenotypes would favour different homozygote and heterozygote states. Selection may also be stabilizing if it favours an intermediate phenotype or disruptive if it disfavours the intermediate and its effect on the frequency of the alleles would depend on the gene actions.

Directional selection experiments seem to indicate that most alleles have partial dominance actions, though many may be involved in multiple locus interactions. Since selection would be expected to favour survival and reproduction of certain genotypes, we expect that it would increase the population level of fitness.

Selection is therefore usually considered to have a positive effect on the evolutionary potential of a population or species in small increments every generation but may not generate an optimal condition at any one time. For alleles that have partial dominance effects, the dynamics of the rates of displacement of one allele by another are such that the frequency of the allele itself affects its own rate of displacement. When alleles are at either low or high frequency, their rates of change are slower than when they are at an intermediate frequency. Except in small populations where accidental loss can occur, selection would not generally lose alleles that are favoured for adaptation.

Another force of evolution is mutation, which is usually considered to be a random change in gene structure and function that occurs indiscriminately throughout the genome. Since most species have evolved a genetic and physiological level that is close to a sufficient capacity for continued evolution, novel changes are not expected to improve performance or fitness but rather to generally decrease adaptability. However, there are always going to be mutations and while most may be disfavoured by selection, there is likely to be a balance between the rate at which new mutants are introduced and the rate of their elimination.

Most mutations occur at low rates for any one locus in any one generation and would not be expected to affect the adaptability of a population unless they are accumulated in small populations. For most conifer species where the frequency of deleterious mutants has been estimated, there are a larger number of lethal alleles carried at one or another locus in most individual trees than in most other plant or animal species. This might indicate that there are many effective loci in trees that can mutate to a harmful state, that the mutation rate is high or that the mutants are favoured in some conditions or are of some advantage in heterozygous conditions. However, independent mutations and selection can generate differences among populations, which can increase fitness in each.

Hybridization among populations can then generate high levels of genetic variance in traits that may then be useful in new environments. A third force of evolution is migration, which can occur among many different kinds of population structures.

The input of pollen or seed from other populations that have different allele frequencies may bring in alleles that might otherwise be lost, and if the migration rate is high enough among all populations may induce all populations to have approximately the same frequencies. If the immigrant alleles are favoured, then further immigration would advance the rate of its increase; if disfavoured, immigration would maintain alleles at higher frequencies than expected in a manner similar to mutation; and if neither, would tend to make frequencies homogeneous among all populations. In the presence of divergent selection among populations that exist in different environments, selection would act to eliminate migrant alleles and would maintain population divergence.

The fourth force is sampling error induced by small population sizes, which tends to increase the probability that low-frequency alleles are lost by chance. If populations remain at low sizes for several generations, the chances of random loss are accumulated; even populations of more than 20-50 adults can lose alleles that may have started with moderate frequencies. Since most forest trees seem to carry many deleterious mutants, even if at low frequency at any one locus, another problem is that small populations will suffer from inbreeding depression and may eventually suffer such debilitation from several mildly depressive mutants that the population goes extinct. Individual populations would have to be large enough to sustain viability, and to avoid environmental accidents and catastrophes if no migrants were recruited.

There are two problems faced in persistently small populations, the loss of alleles and inbreeding depression; hence, larger population sizes are generally better for avoiding extinction. However, the effects of selection, migration and mutation are confounded with the sampling errors incurred in real populations of finite size. Thus, if populations are separated and selected for different environmental optima, migration can introduce alleles in high frequency that depress fitness. On the other hand, migration can maintain alleles in populations where they might be useful, but would be lost due to sampling accidents in small populations.

It is reasonable to expect that all species are under the influence of several or all of the forces of evolution simultaneously and that the balance between the several forces is not the same for all genes in all populations. Furthermore, since the physical and biotic environments of forests are rarely at an equilibrium, the genetic and ecological dynamics of most species would have to be considered to be in other than a stable state.

The actual distribution of the multiple functional forms of alleles at each of the several tens of thousands of gene loci in each of the individuals of a population must therefore be the resultant of the mixture of forces felt at each locus, subject to historical events. The vast possibilities of mixed distributions are overwhelming, although it is also obvious that for most of the species we can study, species are not random mixtures of an infinite number of possible combinations.

11

Social Structure of Farms, Farm Households and Labour

RURAL SOCIOLOGY

During the 1950s and 1960s rural sociology as a whole, and especially rural sociological analyses of agriculture, closely reflected the character of the larger sociological discipline. A distinctive division of labour between sociological theory and empirical research had emerged, with the theoretical enterprise being dominated by Parsonian functionalism and empirical research by quantitative analysis of survey data.

Empirical researchers did seek guidance from the theory community, but they were by and large limited in doing so because the methodological innovations of Paul Lazarsfeld and others were primarily suited to micro units of analysis that were inappropriate for directly testing the macrosociological and macrocultural formulations of Parsons and others. To the extent that there was a linkage between structural-functionalism and quantitative research, it was largely carved out by Robert K. Merton through his notion of "theories of the middle range" and his exemplary collaboration with Paul Lazarsfeld at Columbia University in the sociology of communication, deviance, interpersonal influence, and so on.

Much the same could be said of rural sociology during the 1950s and 1960s. The theoretical and empirical research enterprises in rural sociology were largely insulated from each other, save for middle-range theories such as diffusion-adoption of innovations and Merton- and Lazarsfeld inspired conceptualizations of the processes of interpersonal influence. "General theory" in rural sociology was rarely drawn upon as a direct source of hypotheses to be tested through empirical research, since the macrosociocultural character of these theories rarely lent itself to the elaboration of hypotheses suitable for testing with the new methodological tools being embraced within sociological and rural sociological circles.

The extent to which there was a disjuncture between theory and research in 1950s and 1960s sociology and rural sociology is, of course, a matter of opinion. That many sociologists did (and still would) disagree with such an assessment can be gauged by the sharp reaction to C. Wright Mills' publication of *The Sociological Imagination*.

Nonetheless, it is generally agreed that there came to be a crisis of sorts in American sociology coincident with — and, in many respects, as a result of — the general social ferment of the late 1960s. This social ferment caused many sociologists to question whether prevailing social theory and research had the effect of reinforcing the status quo of racism, militarism, and inequality. Others came to be influenced by theoretical innovations in Europe that were nonfunctionalist, if not antifunctionalist.

These new influences, combined with a rediscovery and reanalysis of the work of many sociologists who had opposed the orthodoxies of functionalism and behaviorism, led to an intellectual ferment that, in a sense, continues today. The late 1960s milieu of challenge to prevailing post-World War II sociological theories and methods culminated in the publication of Gouldner *The Coming Crisis of Western Sociology*.

Rural sociology, however, was not to experience this intellectual ferment until the mid-1970s, and its origins were to be of a somewhat different character than those that shaped sociology at large. Perhaps the most immediate influence on the theoretical and research traditions in rural sociology was the disillusion with prevailing international development efforts that began in the early 1970s. Also, several rural sociologists deeply engaged in international development research became acquainted with theoretical perspectives prevalent in Latin America and other Third World countries that represented critical challenges to Western development orthodoxy. Havens' paper on development theory in *Sociologia Ruralis* was, to the best of our knowledge, the first publication by a North American rural sociologist to confront directly the theoretical and methodological orthodoxies of then-contemporary rural sociology.

Many of the concerns of internationally oriented rural sociologists came to be seen as directly parallel to agriculturally related phenomena in the United States. Critics of the "green revolution," for example, were concerned that these technologies were biased against small-holder peasants, landless agricultural laborers, and the rural poor and were ecologically destructive. So, too, U.S. rural sociologists came to be concerned whether technological change in agriculture was having the effect of marginalizing family farmers and farm workers here at home. Others began to cast a critical eye on the structure of agricultural research and the role of rural sociology in the development and diffusion of new agricultural technologies. Still others came to be concerned about the ecological impacts of modern agriculture and argued that researchers must consider ecological variables if they want to understand the social organization of and technological change in agriculture. Other impetuses to the emerging intellectual ferment in rural sociology included the concerns raised by late- 1960s social movements (especially the environmental, civil rights, and other social justice movements) and the writings of several prominent nonsociologists whose work came to set the agenda for rural social science. Perhaps the single most influential person in this regard was Jim Hightower, who raised provocative questions about agricultural research and the land-grant universities, the effects of agricultural technology on American farmers, the status of the family farm, the rising prominence of large corporate farms, the ecological consequences of agriculture, and so on.

While Havens and his international development colleagues began to develop alternative theoretical and methodological formulations for international rural sociology, William Heffernan and Richard Rodefeld were the first to do so for domestic rural sociology and the sociology of U.S. agriculture. In a sense, Rodefeld's work drew on theoretical postures that had been prevalent in rural sociology for decades.

But while this theoretical posture was seemingly quite conventional, Rodefeld's conclusions were quite at variance with received notions of the benign character of technological and farm structural change. Rodefeld argued that the traditional family farm was undergoing a process of disintegration and was being replaced by corporate-industrial and larger-than-family farms. Rodefeld was extremely influential in rediscovering Goldschmidt's As *You Sow* and in suggesting that the demise of family farming was adversely affecting rural communities. Many of these themes appeared contemporaneously in the work of Heffernan, but Rodefeld's integration of these themes into monograph-length works enabled his formulations to be highly visible inside and outside of rural sociology.

It is important to recognize that the new sociology of agriculture as it emerged during the mid- to late 1970s was theoretically diverse. For example, Rodefeld and Heffernan had their impact by revising fairly traditional theories and demonstrating that seemingly "natural" trends toward structural differentiation in agriculture had adverse implications for family farmers and rural communities. Later there came to be developed a theoretical tradition in the sociology of agriculture that drew heavily on Marxist political economy and, in particular, on the classical treatments of agricultural political economy by Marx, Kautsky, and Lenin.

During the same year the Rodefeld *et al.* volume was published, a set of articles appeared by Mann and Dickinson, Friedmann, and Newby that opened up whole new vistas in the sociological analysis of agriculture through the application of Marxist theory. This trend was consolidated with the appearance of an anthology edited by Buttel and Newby, the publication of a book by Friedland *et al.*, and an anthology by Havens *et al.* More recently, the political economy of agriculture has taken a neoWeberian direction, which was stimulated by Newby, Newby *et al.*, and Mooney. Finally, the post-1980 new sociology of agriculture has been influenced by an ecological posture that was originally codified by Dunlap and Martin.

The new sociology of agriculture has been quite diverse theoretically, but there are several common characteristics of this reorientation of rural sociological research. First, the new sociology of agriculture has been much more ambitious theoretically than the research traditions that prevailed prior to the early 1970s, attempting to combine broad, macrosocial theorizing with the elaboration of falsifiable theoretical formulations and testable hypotheses. Second, the new sociology of agriculture has a greater representation of historical and qualitative research methods than was typical in rural sociological research on agriculture during the 1960s. Finally, the glue that has tied these diverse theoretical approaches together over the past decades has been that these perspectives have tended to be critical, albeit in diverse ways, of prevailing agricultural institutions.

But just as social psychological-behaviorist perspectives that predominated in the 1950s and 1960s did not wholly supplant the community study perspective on agriculture, so, too, the new sociology of agriculture has by no means displaced social psycho logical — in particular, diffusion-adoption — research. Indeed, there have been major revisions of the diffusion-adoption perspective that have enabled it to remain viable and to contribute to the understanding of agriculture. Nonetheless, the neo-Marxist, neoWeberian, and ecological theories have been most influential in shaping the issues of concern to sociologists of agriculture for more than 10 years.

SMALL FARMS AND PART-TIME FARMING

The nature, social bases, and socioeconomic consequences of "small" farms and part-time farming are, in a sense, among the most longstanding foci in rural sociological research on agriculture. Nevertheless, sociological analysis of small-scale agriculture and part-time farming received substantially increased attention and exhibited considerable theoretical reformulation in the 1970s and 1980s for several reasons. First, spearheaded by the writings of Hightower and others on the demise of "small" farms, there was a startling renaissance of concern with the status and prospects of small family farms, which had greatly subsided a few decades earlier in the aftermath of the apparent resolution of the depression-induced increase in tenancy.

Second, quite ironically, several observers began to document the fact that the 1970s had witnessed an unprecedented trend toward the stability and persistence of small farms; in most regions the absolute number of farms with fewer than 50 acres has increased over the past one to two decades. Third, it became apparent that much of the reason for the enhanced viability of small farms was the growing access of farm households to off-farm jobs and income; indeed, the U.S. Department of Agriculture began to note that, in most years, the majority of farm households' aggregate income came from nonfarm sources and that there had been a slow but steady increase in the prevalence of off-farm labour market participation among farm families since World War II.

Small and part-time farms are treated here as a particular focal point in the new sociology of agriculture, a tack that has been taken by many others. Whereas there is a key connection between small-scale agriculture and part-time farming — the fact that part-time farming has in recent decades been much more common among "small" than among "medium-size" or "large" farms — the two phenomena, however, are by no means equivalent. In particular, some "small farms"-which provisionally can be referred to here as low-sales-volume farms, with about $50,000 or less in gross annual sales volume-have little or no off-farm income and, accordingly, have long been overrepresented among rural poverty groups in the United States and elsewhere.

Moreover, in all advanced capitalist societies, especially the UK, part-time farming occurs across the whole spectrum of sizes of family-proprietor farms; in the UK it has been quite common for the operators of very large farms to have off-farm occupations, especially the operation of small businesses related to agriculture. Thus small-scale and part-time farming, although tending to be significantly related, are by no means coterminous. Further, the analysis of neither small nor part-time farming has been dominated or influenced by neo-Marxist postures nearly as much as other areas in the new sociology of agriculture. In the huge literature on these topics there have been theoretical and empirical neo-Marxist works, but these approaches have been a distinct minority component of the literature.

It is useful to begin with a very generalized model of the recent evolution of U.S. agriculture to situate the theoretical and empirical issues relating to small and part-time farms. Over the past decade or so there has been growing acceptance of the notion that U.S. agriculture has followed a dualistic or bimodal pattern of development since the early 1970s. There is evidence that prior to the late 1960s (and thus preceding the phase of dualistic development) there was a singular pattern of increased concentration and differentiation in U.S. agriculture in which there was a direct association between size of farm and its persistence or viability; thus large farms increased in numbers, acreage, assets, and so on at the expense of small farms, particularly small tenant farms without the means to acquire off-farm income. Farm operators with access to off-farm income were thought to be largely in a transitional status, either into or out of agriculture.

It is now suggested that a more complicated or dualistic pattern of concentration and differentiation has emerged. The stylized facts of this association are, first, that large farms — including large family-type, larger-than-family, and industrial farms in Rodefeld's terms — are, much as in previous decades, exhibiting a demonstrable increase in their shares of aggregate production, sales, assets, profits, use of hired labour, and so on.

Second, however, it is argued that small farms are increasing in numbers and holding their own in terms of acreage and aggregate production. This bolstered status of small farms is considered to be made possible largely by access of their operators and other family members to off-farm wage income. U.S. small farmers have been noted to have, on average, total family incomes in excess of those of medium-sized farmers and the U.S. median family income because of the higher than average educational attainments of members of small farm households and their consequent ability to acquire highly remunerative off-farm work. While outside of the South small farmers have tended to have higher educational levels than larger, full-time operators for several decades, the educational disparities between small, part-time and larger, full-time operators are suggested to have increased. This has often been explained in terms of the 1970s "population turnaround" in which well-educated persons, many of urban origin, were attracted to farm residential situations for lifestyle reasons.

Finally, it is asserted that there has been a "disappearing middle" — that medium-size (traditional family-type) farms have decreased in numbers and in shares of aggregate output, assets, profits, and the like. It has been suggested that these medium-size farms enjoy neither the advantages of bigness or of smallness. In particular, "middle farmers," because they have operations too large to be farmed on a part-time basis, tend to lack the labour resources (and often the educational qualifications) to participate effectively in off-farm labour markets. Thus "the middle" is tending to "disappear," albeit slowly. Approximately 5, 70, and 25 percent of U.S. farms are seen to lie, respectively, in the "large," "small," and "medium" categories.

These stylized facts appear to be largely accurate, although Ehrensaft *et al.*, and Barlett have raised empirical and theoretical objections to the disappearing middle thesis. Salamon and Davis-Brown's findings suggest that the "disappearing middle" may have come more from certain segments of the farm population (the more risk-accepting, entrepreneurial farmers) and less from other segments (the "yeoman" farmers whose key goals are preserving their family farms). Similarly, this conclusion may not hold for black farmers whose limited access to credit for land purchases and mechanization led to rapid declines in the total number of black farmers. Nonetheless, there has been general confirmation of this empirical model for the U.S. farm sector, which provides a preliminary framework for placing the phenomena of small and part-time farming in historical context.

As Coughenour and Wimberley have noted, "most American farm families live on small and part-time farms." Thus in the sense of understanding the composition of the U.S. farm sector, it is essential to account for the status of and trends in small-scale and part-time agriculture. A dramatically different context of research on small farms emerged in the mid- to late 1970s. This context can now be seen as contradictory; at the same time that there had been a rekindling of concern with the status and prospects of the small-scale farm that had been unprecedented since the Great Depression, there was a correspondingly unprecedented increase in the numbers of small farms in most regions of the country.

The changing context of the decline and persistence of smallscale agriculture strongly influenced the conduct of research on small farms. There emerged a number of studies examining the factors leading to small farm decline, persistence, survival, adaptation, and so on. The dominant finding in this research was that small farmers pursue multiple strategies for survival — access to off-farm income; minimizing debt; direct marketing of high-value, specialty crops; and so on — and that the choice of strategies was related to the socioeconomic characteristics of the farm, farm operator, and farm household members.

Recognition of the diversity of small farmers and their survival strategies set forth a related line of inquiry: documenting the internal differentiation among small farmers and generating typol ogies of small farms and small farming households. Major studies in this genre include Buttel and Gertler, Schulman *et al.*, Mufioz, Schroeder *et al.*, and Heffernan *et al.*. These studies have typically found that gross sales (or other indicators of scale), age, education, motivations for farming, household composition, off-farm occupation, and off-farm income are important dimensions of internal stratification among small farmers.

A third cluster of research activities relating to small farms has been to explore small farmers' sociopolitical attitudes. One major area of inquiry has been in documenting the motivations of small scale farm operators and their family members.

Another focus of attitudinal research has been on the political orientations of small farmers. Research to date has generally not reported that small farmers have dramatically different political orientations than large farmers on either general dimensions of political ideology or orientations toward agricultural politics. It has been demonstrated, however, that there are major differences in political orientations among small farmers that are primarily accounted for by off-farm occupation, total family income, and race.

Finally, there has been a nagging concern among rural sociological researchers as to how small farms should be defined. Gross farm sales has for a very long time been the principal scale variable used by the Census of Agriculture in collecting and reporting data, and our macro-sociological knowledge base on trends in the structure of small-scale farming is largely based on gross sales data. Nonetheless, it is widely recognized that gross farm sales have significant limitations as a measure of scale or "smallness."

For example, many economists would prefer value-added over gross sales as a measure of scale. But in the absence of a convenient measure of value-added (or some other concept preferable to gross annual sales) for census and survey research purposes, some arbitrary criterion based on gross sales, sometimes combined with a total family income criterion, has continued to be the measurement convention in small-farm research. Nonetheless, as Carlin and Crecink have shown, alternative definitions of small farms have significant implications for the numbers and characteristics of farmers and farm households included in that category.

Part-time farming research in the United States was initiated during the Great Depression of the 1930s, mainly by agricultural economists and to a lesser extent by rural sociologists, and was part of the concern about "agricultural adjustment" that prevailed in USDA and land-grant university circles. It was observed that many farmers were attempting to survive depression conditions and avoid (or extricate themselves from) tenancy by acquiring off-farm income. As was also to become the case in the 1980s, part-time farming came to be widely advocated as a partial solution to the difficult problems of the agricultural depression of the 1930s.

In the decades up until the early and mid-1970s, however, it was recognized that "agricultural recovery," which typically was conceptualized in terms of increased owner-operation of farms, did not correspond to a decrease in part-time farming. Indeed, coincident with agricultural recovery from the late 1930s to the 1960s was an unprecedented exodus of farmers from the land — a new "agricultural adjustment" problem born of expansion, dynamism, and what latter-day neo-Marxists would refer to as differentiation, rather than depression and decline — that again called for part-time farming as an adjustment mechanism.

Whereas in the 1930s farm size and part-time farming did not appear to be strongly correlated, by the 1950s it was widely recognized that there had emerged a strong inverse association of gross farm sales and various indicators of the involvement of the operator in off-farm work. Among many researchers, especially agricultural economists, the principal research concern, on which there was a decade of conflicting results, was with whether part-time farmers were as efficient as full-time farmers.

Whereas rural sociologists were a minority among researchers on part-time farming during this period, Fuguitt in particular made several important contributions to the literature that continue to be drawn upon today. Fuguitt first demonstrated that part-time farming represented a combination of "push-pull" factors — that the prevalence of part-time farming was positively related to off-farm employment opportunities and inversely related to opportunities in agriculture. Fuguitt then established that there were categories of part-time farmers (based on farm and nonfarm origin and on farmers' aspirations) with distinctly different social characteristics.

Fuller has noted that after 1975 research attention to part-time farming increased dramatically and that several new issues occupied the attention of rural sociologists and kindred social scientists. Definitional matters received major attention after 1975; in particular, there came to be dissatisfaction with conceptualizing part-time farming in terms of the presence or absence, or the number of days, of off-farm work by the operator — the way that off-farm employment data have traditionally been collected and reported in census sources. A consensus began to emerge on conceptualizing part-time farming in terms of the activities of the farming household rather than the farm operator. There also continued to be typological research, following Fuguitt's efforts.

Further, there was a significant amount of research on mobility into and out of part-time farming, on the significance of differences in the off-farm occupations that are held, on the implications of part-time farming for rural communities, and especially on the motivations of part-time farmers. Finally, there has emerged a significant political economy tradition in small and part-time farming research that has explored issues such as the rooting of part-time farming in the concentration and differentiation process, the roles that part-time farming plays in accumulation and legitimation in the agricultural and larger economy, the role of off-farm work in the sexual division of farm household labour, and the "embourgeoisiement" thesis — that part-time farming, by enabling smaller farmers to avoid proletarianization, may blunt the emergence of radicalism by subordinate groups in the agrarian class structure.

In light of excellent review articles on the status of part-time farming research, there is a less pressing need to summarize what has been a voluminous literature. Instead, we will comment on some "metathemes" in this literature that have not yet been the subject of thorough reviews.

First, there is a general tendency for the literature on part-time farming to be polarized into highly structural (if not structuralist in the Althusserian-Marxist sense) and highly voluntarist accounts. Structuralist accounts see part-time farming in almost teleological fashion — that the dynamics of capitalism lead inexorably to agrarian proletarianization in (among others) the form of part-time farming, which provides a mechanism for sustaining profitable capital accumulation in farm, and especially nonfarm, industry by providing a low-wage labour reserve. The subjective motivations and experiences of part-time farming households are thus ignored and viewed to be epiphenomena of larger laws of motion of capitalist development. This approach has historically been prominent in Italian rural sociology circles.

Voluntarist accounts, on the other hand, have largely focused on the motivations of members of part-time farming households — especially their desire to remain in farming and to fulfill lifestyle aspirations. To the degree that nonsubjective factors are considered, they are treated as an "opportunity structure" and not as a social structure that is rooted in the historical and material dynamics of agriculture. Fuller has made a related diagnosis of the state of research on part-time farming — namely that despite widespread social science research on part-time farming, as yet there exists no satisfactory explanatory theory.

Fuller attributes this to the fact that there has been disproportionate attention to the "farming" aspects of part-time farming and too little attention to the family and other nonfarming aspects of rural life. We would suggest that the frontier of research in part-time farming should lie in integrating objective and subjective aspects of these phenomena, which should help to stimulate the theoretical thinking that Fuller has found lacking.

A second metatheme that can be distilled from the part-time farming literature concerns the need for a truly comparative approach to part-time farming research. This observation is, in a sense, ironic, since there has probably been more cross-national comparative observation of part-time farming than in any other subject matter in the sociology of agriculture. But while cross-national variations in part-time farming are widely recognized, these divergent systems of parttime farming have yet to be explained theoretically.

This suggests the need for a comparative, integrative perspective on agricultural and rural development, which, like de Janvry's rendering of Lenin's classical schema of "roads of agricultural development," would examine the interrelations between agriculture and the larger, nonfarm rural economy and society. A major step in this theoretical direction has been taken by Bonanno, though his focus has been more on the persistence and role of small farms than on part-time farming per se. Nonetheless, much work remains to be done to explore part-time farming and other aspects of agrarian structure from the vantage point of the dynamics of agrarian and regional-development dualism.

GENDER AND AGRICULTURE

The role of women in U.S. agriculture was by no means a new area of inquiry when interest in this topic soared in the late 1970s and early 1980s. Wilkening, in particular, had sustained a major research programme on the work roles of farm women and men since the early 1950s. And for several decades there had been a substantial amount of attention paid to rural and agricultural family structures. Nonetheless, rural sociological research on gender and agriculture was to embrace distinctly different themes in the era of the "new sociology of agriculture". Whereas there have been a number of foci in this new tradition of research, the dominant cast of roughly 20 recent studies has been to explore the "invisibility" of women in U.S. agriculture, to paraphrase the title of Sachs' recent book-length work in this area.

Research on gender and agriculture has fallen into four major categories, which basically have represented stages in the development of knowledge on the role of women on American farms. The first category is that of cross-sectional descriptive studies that have sought to document the nature and extent of farm women's involvement in on-farm, household, and off-farm tasks. The second is a set of studies that have attempted to assess changes in women's involvement in on-farm, household, and off-farm tasks by utilizing longitudinal research designs. The third category is that of studies exploring the interrelations between the on-farm, household, and off-farm roles of farm men and women. The final category is that of comprehensive theoretical and empirical approaches that attempt to understand how the institutions of capitalism and patriarchy interact to shape the involvement of men and women in various work roles.

The first descriptive phase of research on gender relations in agriculture was initiated by Joyce and Leadley, who summarized the existing literature on rural women, both farm and nonfarm. Joyce and Leadley lamented the lack of research on rural farm and nonfarm women and set forth an agenda for research. Soon thereafter a number of studies would be conducted in which many of the basic parameters of women's participation in farm, off-farm, and household activities were clarified.

Because census data have included little information on farm women and thus have seriously underestimated the economic contribution of women to farming operations, the knowledge base on the productive activities of farm women has been greatest and most longstanding in terms of off-farm employment. It had been recognized for quite some time that farm women were increasing their participation in off-farm labour markets, and that women's off-farm earnings were often crucial in ensuring the viability of farms. It was also recognized, however, that farm women's off-farm jobs typically were concentrated in low-skill, low-wage sectors of the economy and tended to involve less-than-full-time or otherwise intermittent employment.

Subsequent descriptive studies on farm women's work roles focused primarily on their on-farm and household work and its relationships with off-farm employment. One of the most extensive studies was by Jones and Rosenfeld, who used 1980 national survey data to profile farm women's involvement in farm work, off-farm employment, community activities, and several USDA programmes. They found a high incidence of women's involvement in most agricultural production tasks, including those typically considered to be "men's work". Over three-quarters of farm women were found to play a role in bookkeeping and other financial management tasks on the farm. Slightly over 30 percent of the farm women (60 percent of whom worked less than 40 hours per week) reported off-farm employment.

A further major descriptive study was Salant's survey of farm households in 29 rural counties in Mississippi and Tennessee. She concluded that farm women are extensively involved in onand off-farm work; over half of the women reported employment either on-, off-, or both on- and off-farm, about 75 percent of whom worked off-farm. Women's off-farm labour force participation was found to be related to age and family life cycle. The off-farm income earned by farm women made a major contribution to the income of households in the survey — in many cases enabling the family to rise above poverty-level status — despite the fact that farm women's employment was largely in low-wage jobs. Women who worked on-farm primarily did so as unpaid family laborers. Only 3 percent of the women in the sample were principal farm operators. Salant did not, however, find any support for the hypothesis that women who work on-farm tend to do so to free the labour of men in order to acquire more lucrative off-farm jobs.

Another important descriptive study is that of Fassinger and Schwarzweller, which was based on a 1979 sample survey of 124 farm households in Michigan. Fassinger and Schwarzweller found that farm size was positively related to the amount and proportion of time spent by farm men and women in on-farm work. Farm size was inversely related to men's off-farm employment but had no association with women's off-farm employment.

They found that men's time devoted to on-farm work was positively associated with women's on-farm work. Fassinger and Schwarzweller found a very high degree of division of labour by gender for both farm and household tasks. On the farm, women were more likely to assist with financial and recordkeeping aspects of the business and to plant and maintain a vegetable garden. Men, on the other hand, were primarily involved in field operations, the purchase of farm inputs and machinery, and the repair of machinery and farm buildings. The lion's share of household tasks was performed by women regardless of the scale of the farm.

Kalbacher's study was the first comprehensive report profiling the role of women farmers in American agriculture. Kalbacher used 1978 census results and data from a 1979 Farm Finance Survey to present a statistical profile of women farm operators. About 5.2 percent of all U.S. farm operators, or 128,000 farmers, were women in 1978 — representing a near-doubling of the number of female farmers since 1970 and a tripling since 1950. Femaleoperated farms were located disproportionately in the South, and tended more often to be owner-operated, part-time, and smaller than male-operated farms.

Female farm operators were also older and more likely to be in a racial-minority group than their male counterparts. But whereas women farmers were less likely than men to report fanning as their principal occupation, their average off-farm income was significantly less, resulting in a much lower average household income for women farmers, though femaleoperated farms typically had lower debt and more favorable debtasset ratios than did male-operated farms.

A further examination of the role of farm women in the United States was conducted by Tigges and Rosenfeld in their study of differences between men and women among "independent" farmers using a national sample. The independent farmer was defined as "a woman or man who has managerial responsibility for the farm operation and lacks a spouse's regular farm labour". They found that, unlike independent farming men who were similar to men whose wives participated in farming, independent farming women tended to have lower net farm incomes and total family incomes than did other women. Part of these differences were explained by differences in type of farm, class, and life-cycle situation variables. Tigges and Rosenfeld concluded that independent women farmers tended to enter farming by default (as upon the death of a spouse) rather than seeking to enter farming on their own.

After several years of descriptive studies documenting the variety of contributions of women to U.S. farm households, the attention of several researchers shifted to changes in these roles over time. The most important study in this regard has been by Wilkening in which he presented data from a 1978 mail survey of Wisconsin farm families, personal interviews with farm couples completed in 1979, and data from a 1962 study carried out in the same region. The data reveal that families remain the most important source of labour on Wisconsin farms, particularly on dairy farms that dominate in Wisconsin.

Nonetheless, Wilkening found that both farm men and women now contribute less time to the farm operation than they did in 1962. Off-farm employment by one of the spouses was significant in the late 1970s on about one-quarter of the farms, and off-farm work was associated with low levels of gross farm and high levels of total family income, but was unrelated to stage in the family life cycle. While Wilkening found that farm women have not played an increased role in farm production activities since 1962, the data did show that farm women have begun to assume major responsibilities for recordkeeping and other financial aspects of Wisconsin farm operations. Women were reported to be equal partners in making important farm decisions on about half of the farms, a slight increase since 1962. Farm women remained largely responsible for the vast majority of household work during the study period.

Testing part of a theory about the sexual division of labour and authority in farm households, Bokemeier and Garkovich found that farm women's self-concepts relating to their identification with their farms' operations were complex in their sources, including scale of farm, farm background, women's human capital, and family characteristics. They identified four roles for farm women, ranging from little or no involvement and authority to full partnership: homemaker, "ag" helper, business manager, and full "ag" partner.

A quite different study oriented toward assessing change in women's and men's work roles was conducted by Gladwin in North Florida. Gladwin was concerned with exploring an extension of the Boserup hypothesis that the capitalization and intensification of agriculture are inversely associated with the level of participation of women in farm activities. Specifically, Gladwin examined the reverse hypothesis — that a decrease in intensification, or marginalization, of agriculture will be associated with an increase in female labour with survey and census data. Gladwin reported that women are now involved to a greater extent in farm production than in the 1920s and 1930s and that the majority of farm women think of themselves as "farmers" rather than farm wives.

But she also reported that off-farm work has become more important for both men and women on North Florida farms. About 88 percent of the farms were part-time farms with significant off-farm income supporting the farm family. These part-time farms tended to produce crops or livestock requiring relatively little labour. Gladwin concluded that there is an association between labour intensification and the growing participation of women in farm work, since the few highly capitalized farms in the region received comparably less labour from women in the farm family than did the more numerous part-time, less-capitalized farms.

The third category of research on gender and agriculture has been that of assessing the mutual interrelations of men's and women's on-farm and off-farm work and how these associations are affected by the scale of the farm. One such study, by Fassinger and Schwarzweller, found that men's and women's onfarm work were positively associated, and each was positively related to farm size. Also, men's off-farm work was found to be positively associated with farm scale, while women's off-farm work was not related to the size of the farm.

Coughenour and Swanson examined 1979 Kentucky farm survey data to explore the associations between farm size and the work status of farm men and women. They did so, however, in a manner different from the previous literature; whereas previous studies have seen men's and women's work statuses as a function of farm size, Coughenour and Swanson suggest that the opposite direction of causality may be important. They explored this hypothesis by developing a four-category farm-type typology based on men's and women's work statuses. They found that there was a modest association between farm types and size of farm (measured by acreage and gross sales). They explored this association further in terms of "exchanges" between men's and women's onand off-farm labour.

They found, for example, that farms in which the woman has off-farm employment forfeited some of the woman's on-farm labour (and hence its contribution to farm size and sales) but gained in income relative to other farms. They also found that the allocation of men's labour — either on-farm, which was positively associated with farm size and sales, or off-farm, which was inversely associated with farm size and sales — was a stronger determinant of the size structure of farms than was the allocation pattern of women's labour. Nonetheless, they did find that the women's role in farming had significant effects — in particular, that women's off-farm work, which reduced or eliminated her on-farm labour input, tended to be inversely associated with the size and sales position of the farm despite the fact that women's off-farm work provided additional capital for the farm operation.

Buttel and Gillespie examined the interdependencies of average weekly hours of on-farm and off-farm work by farm men and women, the use of hired labour, and the size of farm with 1979 New York farm survey data. They found that farm men and women tend to specialize mutually in either on-farm or off-farm work, with these interdependencies being greatest among small farm households. It was also found that hired labour tends to be substituted for women's on-farm labour input, with this inverse association between hired labour and women's on-farm work being greatest among larger, commercial-scale farms. The results of the ButtelGillespie study were largely corroborated by a replication and extension using data from tobacco-, peanut-, and soybean-producing farms in North Carolina by Simpson *et al.*

The extension pertained to distinguishing between production of commodities that is "organized around" individual units ("small batch") and those whose production is coordinated as a "continuous process". Dairy and tobacco were presented as examples of the former type and soybeans and peanuts as examples of the latter type. Simpson *et al.* found that crop production organized as continuous processes tended to make farm women's work more independent of farm men's than was the case for crop production organized around small batches, and thus technological change in the production of "continuous process" crops had different effects on men and women.

Rigorous analysis of gender relations on U.S. farms is a sufficiently new focus of research that, save for the pioneering book by Sachs, there has been little effort to incorporate gender relations phenomena into a comprehensive theory of farm structural change. Sachs' book thus is the single representative in the fourth category of research on farm gender relations and the gender division of farm household labour.

Sachs' book is a wide-ranging treatment of research on farm women from a variety of disciplines and in both First World and Third World contexts. For present purposes, Sachs' treatment of U.S. farm women receives principal emphasis. As indicated in the title of her book, one of Sachs' key arguments is that while women have historically made significant contributions to agricultural (and, increasingly, off-farm) labour, these contributions have been overlooked and undervalued.

Whereas Sachs roots her analysis in the new political economy tradition referred to at the beginning of this chapter, she contends that the invisibility of the women's role in agriculture, and the subordination of women in the agricultural labour process, cannot be explained with reference to the dynamics of capitalist development alone. She argues that the simultaneous contribution of women to agriculture along with their invisibility and subordination must be viewed through the dynamics of capitalism and patriarchy, which is defined as a "set of social relations which has a material base and in which there are hierarchical relations between men, and solidarity among them, which enable them to control women". She emphasizes, moreover, that patriarchical relations preceded (although they have been transformed by) capitalism and thus cannot be seen as a direct consequence of capitalist development.

Sachs begins by demonstrating that at each major stage of agricultural development from the seventeenth century to the present women have played important roles in facilitating the viability of U.S. farms. Until the twentieth century the importance of women was typically seen to be in producing and processing food for home consumption. There was a traditional division of labour between men and women in which men did the majority of the field work, but this division of labour would often break down during busy seasons when women would do "male" tasks. With the industrialization of U.S. society and further commercialization of agriculture, women's subsistence production activities diminished in importance. Women accordingly tended to increase their on-farm and/or off-farm work. These patterns, however, varied substantially by region, with women's direct involvement in farming being greatest in the South and West. Sachs notes that one constant historical feature of women's roles in farm households has been in their performing the bulk of household tasks. Women have become somewhat more involved in farm decision making, but Sachs suggests that their decision-making roles have been secondary to their labour contributions to the farm operation and household.

Sachs' analysis then turns to the historical development of a "domestic ideology" — that women are naturally "domestic" beings and more suited to household work than men — in U.S. society in general and in agriculture in particular. She argues that keeping women in domestic roles has facilitated the development of capitalism by maintaining a reserve labour force and creating a consumptive domestic class. Sachs suggests that the family farm, particularly because it is based on the nuclear family, has been one of the most pervasive mechanisms through which patriarchal relations have been maintained on U.S. farms.

She then explores how this patriarchal order has been reflected in the sexual division of labour on farms and in the social relations of women-operated farms. Sachs' book has been exemplary in identifying how gender relations on U.S. farms have been shaped by the twin forces of patriarchy and capitalist development. Her work remains the most comprehensive theoretical statement and overview of empirical evidence on farm gender relations that has yet appeared.

Sachs' book is wide-ranging but brief, suggesting that further elaborations and critiques will increase our theoretical and empirical understanding of farm gender relations and the gender division of farm and rural labour. In particular, several ambiguities remain in the literature. At an empirical level, while the past decade of research has established a number of frequently supported generalizations about the structuring of farm women's work roles, there are a number of conflicting results.

Some observers, for example, argue that women make major contributions to farm work, whereas others see a clear division of labour such that women participate very little in farm tasks. Other empirical controversies may well arise from the fact that most of the major studies in the field have been based on data from a single state.

Thus further research is needed to clarify these empirical patterns and to incorporate them in a larger theoretical perspective — a major component of which will likely revolve around establishing and explaining differential patterns of women's work roles across farming regions. Some of this theorizing and comparative research is, however, taking place in Europe, for example, has analyzed in detail the gender relations on farms in France and has made some initial notes toward a theory of variations in gender differences in the household, child-rearing, and farm work.

POLITICAL ECONOMY AND THE NEW SOCIOLOGY OF AGRICULTURE

The most distinctive aspect of the new sociology of agriculture in the United States has been a strong representation of and legitimacy accorded to Marxist and neo-Marxist perspectives. This has been unprecedented in U.S. rural sociology (despite the fact that Galeski, 1972, and others had developed comprehensive neo-Marxist treatments of Western agriculture well before the rise of the "new sociology of agriculture" in North America). Perhaps Steeves' article in *Rural Sociology* was the first instance in which RSS' official journal published a paper based largely on Marxist theory. It was not, however, until the late 1970s that there began to be elaborated systematic Marxist explanations of the dynamics of American agriculture by U.S. scholars.

Pioneering papers in this tradition were prepared by Mann and Dickinson, Friedmann, and Newby. The Mann-Dickinson paper and one of the Friedmann papers were published in the *Journal of Peasant Studies*, a British journal that had been the publishing vanguard of a revitalized political economy of peasant studies and agrarian history in Europe. The third major contribution to the development of a Marxist political economy of agriculture, that by Newby, was somewhat idiosyncratic in the sense that Newby was and still is one of the most influential "neo-Weberians" in the United Kingdom.

In Newby's 1978 paper he nonetheless made a strong case that the work of the German Social Democratic Party theorist of the turn of the century, Karl Kautsky, especially his book *Die Agrarfrage*, had palpable relevance for understanding the structural dynamics of American agriculture. Newby has elaborated this argument in a further paper in which he has suggested that a fruitful sociology of agriculture must be based upon an integration of the perspectives of Kautsky, Marx, and Weber.

What is, in retrospect, remarkable about this early phase in the development of a Marxist political economy of agriculture is that the earliest contributors to this line of scholarship had to stray very little from the classics in political economy. The Mann and Dickinson paper, for example, drew fairly directly from Marx *Capital and Grundrisse* and secondarily on Lenin's work. Friedmann likewise drew on the work of Mandel, especially his *Marxist Economic Theory*. Newby, focused largely on Kautsky's work. One should not ignore the creativity of their reformulations of these classical ideas, but there had nonetheless been a long classical tradition in the Marxist political economy of Western agriculture.

It is useful to begin this chapter by recalling why, aside from questions of scholarly acceptability, Marx had been almost totally ignored in analyses of changing agricultural structures by North American rural sociologists until the late 1970s. In addition to the unpopularity of Marxian ideas in the land-grant system, the key reason for the neglect of his work in the study of agriculture probably has been that Marx's stylized model of the polarization of economic enterprises (according to the laws of the centralization and concentration of capital and of proletarianization) into the antagonistic classes of capital and labour seemed to be negated by the very persistence of the family farm.

Marxist sociologists of agriculture, however, came to question this dismissal of Marx in two different ways, which continue to represent the major axis of debate and orientations toward research within this emergent Marxist tradition. On one hand, Mann-Dickinson, Friedmann, and others have rooted their analyses in elaborations of Marx's own arguments about why one might predict that the particularities of agriculture as a production sector would cause agriculture to experience far slower and more uneven capitalist development than would other branches of industry. Newby and, more recently, de Janvry, Goss *et al.*, Friedland *et al.*, and others have, on the other hand, raised arguments that there has been and will continue to be demonstrable capitalist development in Western agriculture, as typified by industrial agriculture in the U.S. Sunbelt and in large-scale farming in eastern England.

CAPITALIST TRANSFORMATION OF AGRICULTURE

Mann and Dickinson's pathbreaking article begins by identifying the weaknesses of prevailing subjectivist arguments as to why family farming has persisted in advanced capitalist societies (though the degree to which Chayanov eschewed structural analysis has often been exaggerated). They suggest, instead, that there are within Marx's work the major elements of a nonsubjectivist or nonvoluntarist explanation of why capitalist development, which they largely define in terms of proletarianization and the establishment of the capital-labour relation at the point of agricultural production, should proceed more slowly in agriculture than nonfarm industry.

They emphasize how agriculture, because of its seasonality, tends to involve a disjuncture (or "nonidentity") between "production time" and "labour time," which creates a barrier to the routinization of the labour process around the calendar and makes agriculture less profitable than other branches of industry. Mann and Dickinson make the case that agriculture is a unique industry in that the seasonal cycle results in interruptions in the use of labour in the production process while the growing commodity is left to natural vagaries.

Thus labour by those working in agriculture ("living labour") is both discontinuous and makes only a modest contribution to the overall production process. Since, in Marxist terms, only living labour creates surplus value, value is neither created nor transferred during these interruptions, thereby creating the nonidentity between production time and labour time. Hence, agriculture tends to be unprofitable and therefore relegated to family labour (or petty commodity) producers. Mann and Dickinson also suggest that agricultural commodities involve longer turnover times (production time plus circulation time [the time required to sell the commodity]) than do other industries, further reinforcing the relegation of farming to noncapitalist producers. Finally, Mann and Dickinson note that agriculture tends to involve producing perishable commodities, which increases the risk of production and therefore makes it even more unattractive to capitalists.

Mann and Dickinson, however, do not claim that capitalist development is precluded in agriculture. They emphasize, in particular, that agricultural research may have the effect of reducing or eliminating the nonidentity of production time and labour time, reducing turnover time, and minimizing the perishability of agricultural commodities. They point out that there are instances of capitalist agriculture in commodities in which a combination of research advances and favorable agroclimatic circumstances have permitted the establishment of capitalist relations of production.

Friedmann's work, although it has taken much the same approach as Mann and Dickinson's, has proceeded along somewhat different explanatory lines. Friedmann has drawn on the Marxist theoretical tradition, but her explanation of the persistence of household forms of agricultural production (which she terms simple commodity production) rests largely on how family farms can meet the competition of — and often outcompete — capitalist farms in the hostile context of competitive markets in means of production and agricultural commodities.

Friedmann begins in much the same way as Mann and Dickinson, pointing toward the high degree of risk and the cyclical demand for labour in most agricultural commodity systems. She argues, however, that the crucial aspect of household forms of agricultural production is that simple commodity producers, unlike capitalists, do not have to earn a profit in order to reproduce their enterprises, *i.e.*, remain in business. Simple commodity producers need only accomplish "simple reproduction." Capitalists, on the other hand, are forced by the logic of competition to strive to earn the average rate of profit, lest their firms become marginalized and eventually be forced out of business.

Moreover, Friedmann has maintained that simple commodity producers in agriculture have a far greater flexibility than do capitalists in reducing their consumption to the subsistence level in order to survive severe market downturns. Capitalist farmers will typically tend to liquidate their assets if agricultural production is no longer able to generate the average rate of profit. Friedmann tested this proposition empirically with historical data which show that the world-market downturn in the price of wheat in the late 1800s led to family producers in the United States and other white settler colonies being able to out compete capitalist producers in England, Prussia, and the United States.

Finally, Friedmann has noted, much as did Mann and Dickinson, that there are "transformational tendencies" in simple commodity production, among them the subordination of independent producers by nonfarm capital. Thus, whereas Friedmann emphasizes the particularities of agriculture that lead to the persistence of simple commodity production in advanced capitalism, she recognizes that there are conditions that may lead to its transformation toward capitalist forms.

Along a similar line and influenced by the world systems school of political economy, McMichael analyzed the cotton plantation system of the Antebellum South, finding that this system was in transition during the period as its basis in mercantile capitalism became transformed into industrial capitalism with an accompanying need for specialized raw materials (cotton).

The credit system that developed concomitantly transformed the southern plantation into a system of fully commercialized relations with all the preconditions of industrial agriculture, save for the presence of slave labour. The lack of "free" wage labour — that is, the reliance of plantations on slavery and later on sharecropping as means of labour recruitment and control — proved to be extremely significant, especially by inhibiting the technological transformation of production. McMichael concluded that understanding the transformation of the relations of agricultural production requires attending to the emerging forms of the capitalist economy, "including political relations, competitive market structures, and the organization of credit institutions". Related work by Gary Green and Zey-Ferrell and McIntosh also suggests that the structural organization of the banking industry affects access to credit by farmers with farms of different sizes, and that this has implications for change in the structure and character of farms. Simultaneously with the publication of the work of Mann-Dickinson and Friedmann, there emerged a very different neo-Marxist tradition, which stressed that U.S. and Western agriculture was rapidly going down the road of capitalist development and proletarianization — a theme to which Marx had given considerable emphasis in major portions of his work. De Janvry, an agricultural economist by training and vocation but a scholar with strong ties to the sociological community, has been among the most outspoken on the issue of the demise of the family farm in advanced capitalism.

Following Kautsky *Die Agrarfrage* and Lenin and building on his own previous work on Latin American agriculture, de Janvry argued that the development of late capitalism has witnessed a startling rate of destruction of family farms and that the forces that now affect agricultural producers — rapid technological change, state subsidies of research and capital investment, and state commodity programmes — make it quite implausible that the family farm can survive.

De Janvry has argued that whereas agriculture tends to experience capitalist development less rapidly than other branches of industry, it is quite likely that the forces of proletarianization and state-subsidized capital accumulation in agriculture will continue to slowly but surely erode the position of the family farm and lead to its differentiation into antagonistic social classes. Thus, in de Janvry's view, independent producers are a transitional class in advanced capitalism; capitalist relations must inevitably and irreversibly penetrate family farming and lead to its demise, much as has occurred in other branches of industry in advanced capitalist societies. Newby, largely writing on behalf of the neglected works of Kautsky, made many of the same points as did de Janvry. It is useful to note in this regard, however, that Kautsky *Die Agrarfrage* contained a series of sophisticated arguments about why there would be a *slow* pace of capitalist penetration of agriculture.

Kautsky argued that capitalist penetration of agriculture, despite its slowness and unevenness, would ultimately proceed and would result in the decomposition of the German peasantry. Kautsky's work can be seen to at once recognize reasons for persistence of the peasantry, yet deny that the peasantry would survive over the long term in advanced capitalism.

The work of Friedland *et al.* has also been a highly visible contribution to the Kautsky and Lenin tradition, which emphasizes the primacy of analyzing emerging capital-labour relations in agriculture and the ultimate separation of independent producers from their means of production. Drawing on Marxist scholarship in the sociology of work and industry as well as on Kautsky and Lenin, Friedland *et al.* developed a comprehensive theoretical position on agricultural development and the selective industrialization of particular agricultural commodity sectors.

While their *Manufacturing Green Gold* book was based on the California lettuce industry, Friedland and colleagues have explored several other commodity sectors in California agriculture, including tomatoes and grapes-raisins. They have given particular emphasis to the penetration of capitalist relations of production into agriculture, but have also stressed that the nature and pace of this penetration vary widely depending upon the commodity system in question. Nonetheless, they take a posture similar to that of de Janvry — that while agriculture in general and particular commodity sectors exhibit a slow pace of capitalist penetration, capitalist relations do appear increasingly and bear many similarities to production relations in other branches of industry.

Agricultural Production Forms into the Capitalist Political Economy

Another thrust in the neo-Marxian political economy literature, and one that is somewhat contrary to those discussed above, has been the argument that the differentiation of agriculturalists into the capitalist and working classes may be incomplete in the foreseeable future as farm and nonfarm production become integrated into a single system incorporating different organizational forms of production. An early statement of this line of thought was that of the German Social Democrat and Marxist theoretician, Karl Kautsky. Kautsky argued that what was crucial to understanding the evolution of agriculture in advanced industrial societies was not simply the dominant form of ownership of agricultural enterprises, but rather the functions that were served by the emerging organizational formsof agricultural production.

This approach was developed further by Mottura and Pugliesi in an historical analysis of small holdings farmed part-time in southern Italy and the functions of smallholder agriculture in contemporary decentralized economic development. The thrust of their argument was that while most agricultural production took place on farms organized along capitalistic lines, part-time farming served as a backup alternative for the workers in industrial plants located in rural areas. In times of industrial contraction and high unemployment, displaced workers with small farms could turn temporarily to subsistence production until industrial conditions improved, thus forming a reserve labour force.

This integration of agricultural and nonagricultural production spheres has been further elaborated by Bonanno, who examined the role of the state in fostering small farms as one strategy to mediate the interests of the conflicting social classes in the emerging social orders of the advanced societies, particularly Italy and the United States. In this light, various farm programmes to deal with problems in the agricultural sector can be seen in part as rooted in the legitimation function served by the continuation of small farms.

Small farms are also important in the decentralizing industrial system fostered by state policy. In this system, industrial firms move to rural areas where labour is not unionized and where wages are low because many potential workers have small farms producing inadequate incomes and few alternative opportunities. Work is also increasingly "informalized" in cottage industry, piecework arrangements. In such contexts, small farms serve the function as "keeper of surplus labour" — providing, at the same time, a source of low-cost labour for industry and, in the face of the tenuous employment, a source of security for the members of households with small farms.

Wenger and Buck, building on this line of thought but especially on the earlier work of Andre Gunder Frank, take this argument another step by examining how exploitation and superexploitation (extracting more value from workers' labour than permits reproduction of that labour) from members of farm households are necessary and dynamic features of both advanced capitalist societies and developing societies. Production organized according to obligations of kinship ("domestic relations of production") link ("articulate") in various ways with production organized along capitalist lines so that value is transferred from the domestic sphere of production to the capitalist sphere of production.

This takes place in such a way as to make the domestic sphere an "interstitial domestic reserve of labour" that subsidizes the capitalist sphere either directly or indirectly through various mechanisms. For example, unpaid household labour reduces both the wages necessary for workers employed in industry and the prices of agricultural commodities required by farm families. Off-farm income from wage work helps to pay the costs of agricultural production and thus lowers the price of food for other working class families.

Farmers as Actors in a Capitalist Political Economy

Whereas the perspectives of Mann-Dickinson and Friedmann on one hand, and of de Janvry, Friedland *et al.*, and others working in the Lenin (and, to a lesser extent, the Kautsky) traditions on the other, are, in a sense, in diametrical opposition, much of the most provocative work in the Marxist tradition in the "new sociology of agriculture" has revolved around formulations that explicitly or implicitly are attempts at synthesis. One of the most noteworthy attempts has been by Mooney.

Mooney's principal contribution has been to cast doubt as to whether the existence of conventional capital-labour relations on farms is an adequate benchmark for gauging the existence of capitalist penetration of agriculture. Following Wright, Mooney has developed a model of agrarian class structure involving "contradictory class locations" such that class positions other than family labour farmer (unity of capital and labour in the farm household), capitalist farmer, and agricultural wage labour are seen to exist. In particular, Mooney sees that there are several "detours" that can be taken by farmers in order to avoid proletarianization.

These detours involve tenancy, contract farming, part-time farming, and debt. In each, whereas there is no capital-labour relation at the point of agricultural production, farmers are exploited by some fraction of nonagricultural capital (in tenancy, by landlords; in contract farming, by agribusiness; in part-time farming, by off farm capitalists; and in debt, by finance capital). Thus, Mooney argues that the exploitation of farm wage workers by agrarian capitalists is only one form that capitalist penetration of agriculture can take.

Moreover, Mooney sees that these detours — that is, the alternative ways in which capital acts to "strip simple commodity producers of... surplus value other than through the extension of wage labour" — may be more significant than full-blown capital-labour relations at the point of agricultural production. It is significant, however, that Mooney's explanation of why these contradictory class locations in agriculture tend to emerge has a subjectivist component, based on Weber's distinction between formal and substantive rationality. Mooney sees that many farmers are motivated more by forms of substantive rationality than they are by formal capitalist rationality. Accordingly, these farmers tend to be tenacious in holding onto their farms and farm lifestyles and will often tend to take one of the four "detours" to capitalist development in order to remain in agriculture.

Mann and Dickinson, however, replied vigorously to Mooney's neo-Weberianism with two major arguments: first, that so-called contradictory class locations are really based on the same social relations of production grounds that Mann and Dickson use and, second, that Mooney has misconstrued their notion of "obstacles to capitalist development" in agriculture to be immutable barriers, so that Mooney was unable to recognize that his notion of "detours" to capitalist penetration of agriculture is similar to their own views.

Mann and Dickinson thus argue that subjectivism, and the explanatory ambiguities it involves, does not yield insights beyond those afforded by a structural, neo-Marxist theory. They were also critical of Mooney's project of synthesizing Marxian and Weberian approaches on the ground that these approaches are incompatible, except on an ad hoc, eclectic basis. Mooney added further to the debate by arguing that Mann and Dickinson have exaggerated his departure from a neo-Marxist framework-that his perspective retains Marxist insights by incorporating key Marxian categories as ideal-types within a Weberian framework.

He argued that this approach overcomes the incompatibilities of the Marxian and Weberian frameworks — unless, of course, one adopts a mechanistic Marxist or Leninist perspective. His key position was that the presumption of a predictably patterned and necessary differentiation of farmers into the capitalist class and proletariat is erroneous and that subjectively meaningful human action affects the process of differentiation substantially.

With the 1980s farm crisis coming under increasing scrutiny by rural sociologists, Mooney has extended his analytical framework to an historical examination of the state-sponsored farm credit system (Farmers Home Administration [FMHA], Commodity Credit Corporation [CCC], and the Farm Credit System [FCS]) and tax policy incentives for increasing the capitalization of agriculture. Mooney conceptualizes credit and tax policies as means of dealing with the crisis of legitimacy posed by the increasing prevalence of tenant farming produced by the Great Depression. Mooney argues that this credit and policy system created more indebted farmers who ultimately became vulnerable to financial crises in the form of rising real interest rates and declining agricultural commodity prices, a situation such as that which occurred in the 1980s. Financial crises expose to perceptive actors the role of the state in creating the crisis, thus creating another crisis of legitimacy and providing a basis for affected farmers to mobilize against the state.

Pfeffer has taken a considerably different tack than Mooney, though Pfeffer's work has some similarities to that of Mooney in that he stakes out middle ground on the question of whether the predominant feature of agriculture in advanced capitalism is the persistence of the family labour farm on one hand, or its demise into capitalist relations of production on the other.

Pfeffer conducted an historical analysis of three systems of agricultural production in the United States (industrial agriculture in California, family farming in the Great Plains, and the sharecropping system in the South prior to World War II) and demonstrated that recruitment and maintenance of access to a suitable labour force was a crucial factor shaping the disparate organizational structures of these three systems. Only in California, largely because of preexisting land concentration and of state actions to recruit and restrict the residential, educational, and occupational mobility of ethnic minority farm workers, was capitalist agriculture able to set root prior to World War II. Noting that the U.S. South had a high degree of land concentration during the entirety of the nineteenth century, Pfeffer goes on to explain why southern planters were unable to recruit a wage labour force after the Civil War and why the tying of sharecroppers to plots on the landlord's plantation through the crop-lien system was the only possible route to recruiting the labour necessary to operate a large plantation.

Finally, Pfeffer demonstrates that a combination of labour scarcity, the unwillingness of immigrants to work for wages on farms, and severe declines in the price of wheat in the late nineteenth century led to the establishment and persistence of family farming in the Great Plains. Pfeffer thus suggests that while capitalist agriculture can emerge under particular circumstances, there have been several alternative forms of agricultural organization — sharecropping and family farming, in particular — that have become established due to regional and commodity variations in the ability of farmers to recruit a dependable labour force.

Still a different approach was taken by Whatmore *et al.*, who have developed a theoretically based typology of farms in Britain. Their goal was to understand the effects of external pressures by nonfamily and nonfarm capitals to penetrate agriculture and how these pressures articulated with internal processes on farms.

By cross-classifying levels or degrees of subsumption of internal and external relations, they developed a typology with four main categories of farm types, ranging from relatively unsubsumed ("marginal closed unit") to the "subsumed unit." The internal relations variable pertained to ownership and control of farm capital and land, control over management, and the balance of family and hired labour.

The external relations variable pertained to level of dependence on industrial capitals for production inputs and purchase of farm outputs and to the degree of involvement in credit relations, especially with finance capitals. Applying this theoretically based typology to the data from three areas in southern England — an urban fringe area, an agricultural area undergoing commercial and industrial development, and a primarily agricultural area — they found that the process of subsumption of farm production relations, and thus their integration into the "wider circuits of capital," varied over time and space. Whatmore *et al.* concluded that farm families need to be understood as actors in the process of subsumption rather than its passive victims.

The neo-Marxian theoretical enterprise (and that of the modernization school as well) on the conversion of traditional peasant economic relations to commodity relations under capitalism has also been scrutinized by Vandergeest.

Among his major criticisms were that neo-Marxist approaches tend to be too unilinear, too macro-structural, too denying of human agency, too inattentive to the role of the state in the process, and too unconnected with practice. He has argued, from a neo-Weberian position rooted in the work of Pierre Bourdieu, that all categories and theories are in the last analysis historically contingent, ideological, and interpretive social products.

The world is a complex whole which can only be investigated empirically and understood through theory. It cannot be reduced to the simple working out of a model derived through deductive theory — whether that of "simple commodity production" or of "capitalism". There is not a single deductive logic (such as the logic of exchange-value or accumulation) underlying or determining all relations in a capitalist formation, but there are different, historically contingent principles which we can only investigate through empirical research.

There has recently been a tendency in the neo-Marxist and neo-Weberian literatures toward a deemphasis on "deductivist"-functionalist theoretical postures. Subjectivist perspectives such as those of Vandergeest and Mooney have been influential in leading to this reorientation, though this tendency is by no means limited to those persuaded by neo-Weberianism. The three postures in the Marxist political economy of agricultural tradition — those emphasizing the persistence of petty commodity production, those emphasizing the inevitable differentiation of petty commodity producers into antagonistic social classes, and those seeking synthetic positions — have yielded a continuing, lively debate in the rural sociology literature. Particularly instructive is the debate over the Mann-Dickinson hypothesis.

Also of importance is Goodman and Redclift's commentary on the work of Friedmann in which Goodman and Redclift critique Friedmann for taking an overly deductive approach to petty commodity production that ignores the role of historical conjuncture and ideology. Bernstein has made a comparable argument — that generic theories of simple commodity production under advanced capitalism tend to overemphasize its "functional" aspects and downplay contradictions and the diversity of household forms of production.

It should be emphasized that theoretical and empirical treatments of the political economy of North American agriculture have not been made only by rural sociologists. Many of the pioneers in this tradition of scholarship have been nonsociologists or sociologists who have had little or no connection with the Rural Sociological Society or the institutionalized form of rural sociology in U.S. land-grant universities. Nonetheless, rural sociologists have made influential contributions to this literature.

The emerging political economy tradition in the sociology of agriculture has led to a provocative and stimulating literature. It should be mentioned, however, that those working from this perspective are just beginning to establish a distinctive research programme; much of the existing literature, in fact, often tends to involve superimposing a new vocabulary on already-established data, such as Goss *et al.* set out to do. Establishment of a distinctive research tradition will probably occur slowly since conventional data sources such as censuses and sample surveys are often inapplicable to theoretical issues in the political economy of agriculture. There have, nonetheless, been several research papers in the political economy tradition that have utilized innovative new methodologies to generate research findings that bear directly on specific hypotheses.

Subculture and Agricultural Structure

It is useful to note that the growing interest in subjectivist approaches in the new sociology of agriculture has not been confined to those who work within the neo-Marxist and neoWeberian traditions. Anthropologically-oriented researchers have also contributed to this literature through the examination of the effects of ethnic background on state and local level farm structure.

Salamon and associates have done the pathbreaking studies in the anthropological tradition in Illinois communities, though recently this work has been taken up by others.

While their work was not intended to contribute to the "structure versus agency" debate that has raged between Mann and Dickinson on one hand, and Mooney on the other, and actually predated much of this debate, it is clearly relevant to the debate. In particular, the work of Salamon and associates pertains to the role of subjective motivations in farm decision making, especially motivations conditioned by persisting subcultural variations based in differences in ethnic origins of farm families.

The key finding of this line of research has been that farm families with German ethnic backgrounds tend to view farming as a way of life and hold a strong value for keeping the family farm intact. These Salamon has labeled "yeoman" farmers.

In contrast, farm families with British ethnic backgrounds tend to be more entrepreneurially oriented, viewing farming as a way to make profits and having little attachment to farming or to particular farms. These Yankee farmers are more likely to seek growth in scale and less likely to support their local communities. This research has led to understanding differentials in farm size as a partial function of differences between these subcultural variants. Flora and Stitz found that yeoman farmers in a Kansas county generally expanded less than did Yankee farmers. Foster *et al.* also found support for the Salamon hypothesis in a study of a sample of Illinois farms that had been in the same families for 100 years or more.

Several other anthropological studies in the tradition of Salamon and colleagues have been reported in Chibnik. Of particular importance is Barlett's article in which she criticizes notions of the demise of the family farm and of the "disappearing middle" from an anthropological perspective.

THE AGRICULTURAL WAGE LABOUR FORCE

Much of the attention of sociologists who have researched industrial agriculture has been focused on issues relating to the hired agricultural labour force. In this book, however, we treat the agricultural wage labour force as a distinct topic, since the bulk of this labour force in the United States (in terms of hours of annual labour) is white, works full-time, and works outside of the industrial agricultural sector. In this chapter we make some observations on research into the structure of the hired agricultural labour force in both industrial and nonindustrial agriculture.

It is useful to make several preliminary points about the U.S. agricultural wage labour force before proceeding with an overview of the relevant literature. First, as Friedland has noted, federal government data sources on the structure of the hired agricultural work force are notoriously inadequate, so that one cannot even develop a satisfactory profile of the socioeconomic characteristics of the employers of these workers. Second, there are many popular misconceptions about the hired agricultural labour force — presumably, at least in part, because federal data sources are so poor. The vast majority of hired workers are white and work full-time (rather than being largely migratory Hispanics and blacks).

Further, the proportion of total hours of labour in agriculture accounted for by hired workers has either been stable (at about 25-27 percent from the turn of the century to the early 1970s) or increased (to about 35 percent by the early 1980s); thus technological change has been much more effective than often realized in displacing family labour than it has in displacing hired workers.

Perry and Friedland have, in fact, noted that trends in agriculture toward concentration and very large, specialized farms may well lead to increases in the magnitude of the agricultural wage labour force despite the displacement effects of agricultural mechanization.

Friedland has reported data showing that, in the 1970s, each state, with the single exception of Connecticut, experienced an increase in the number of hired workers per farm. Third, our knowledge of the structure of the wage labour force in industrial agriculture is far superior to that for the labour force of family-proprietor farmers.

Because several useful summaries of the literature on the wage labour force in industrial agriculture exist, we make only a few general comments here. Thomas-Lycklama a Niejholt, a Dutch sociologist, prepared a useful study of the East Coast migrant labour stream — one of the three major such streams along with the mid-continent and West Coast streams — based on survey data collected among hired workers in 1970 and on documentary data. She has made several observations that parallel those made by Friedland *et al.*, Thomas, and others who have studied farm labour in the Southwest. First, the key element in migrant agricultural labour markets tends not to be labour sufficiency (*i.e.*, enough labour to meet demand), but rather labour surplus, which creates cheap labour.

Second, agricultural mechanization, whereas it has historically decreased aggregate demand for migratory labour, has not eliminated this demand; in particular, mechanization, since it is typically accompanied by specialization, tends to result in seasonal bulges in labour demand and continued dependence on abundant, cheap, mobile labour. Mechanization thus creates demand for a large number of workers to be employed for very brief periods of time. Some large growers, however, have shifted from an "external" labour force, dominated by migrant workers, to a more permanent, "internalized" labour force where agroclimatic and technical conditions permit.

Third, migrant workers tend to migrate much shorter distances than often assumed; of the workers studied by Thomas-Lycklama a Niejholt, about 50 percent worked in only one state or two, rather than migrating incrementally along the entire East Coast migratory stream from Florida to the Northeast. Fourth, she reported that migratory farm workers are not uniformly disadvantaged. There are substantial income differentials, which are largely explained by ethnic-racial status. Mexicans were the most disadvantaged, followed by West Indians, Puerto Ricans, blacks, and whites.

She also found that migratory workers, who are predominantly nonwhite, tend to receive lower wages than nonmigratory workers, who are predominantly white.

It should be noted, however, that Thomas has reported evidence that, in the lettuce harvest wage labour force in California, citizenship and gender tend to be the principal bases of wage differentials, with noncitizens and women receiving lower wages than citizens and males. Thomas has stressed that subordinate status in terms of citizenship and gender are resources utilized by employers to discipline the lettuce harvest labour force.

Thomas, Friedland *et al.*, and Majka and Majka along with Thomas-Lycklama a Niejholt have also stressed the role of state policy in facilitating labour surplus, curbing the establishment and militancy of farm labour unions, and permitting the payment of relatively low wages. As suggested earlier, while there have been hundreds of studies by sociologists and nonsociologists of the hired labour force in industrial agriculture, little attention has been given to the largely permanent, full-time, white workers recruited by family-proprietor farms from the community or surrounding region in the nonindustrial agricultural regions.

There have, however, been three such studies — one in England by Newby and two in the United States — that provide an encouraging basis for future research in this area.

Newby's studies of petty capitalist agriculture in England began with several theoretical and empirical issues relating to the relations between large farmers and their hired workers. Whereas farm workers make relatively low wages and perform demanding work, they nonetheless tend not to be politically radical, to be dissatisfied with their work, to join labour unions, or to otherwise dissent from the discipline and authority of their employers. Newby explored the role of paternalistic controls in enabling capitalist farmers to maintain a relatively docile labour force despite low wages and low levels of living. He has emphasized the fact that farmers tend to recruit workers who have agricultural backgrounds and are from the local area, thereby reducing the impersonality that typically exists in urban-industrial labour markets.

Employers also tend to provide part of their workers' compensation in the form of housing on the farm, which further ties the worker to the employer. Finally, Newby observed a close correspondence between workers' sociopolitical values and images of the class structure and those of their farmer employers. He suggests that workers have internalized many of the ideological constructions of their employers and have remained deferential rather than rebellious toward them.

Barlett, largely following Newby's research design, has studied the full-time hired labour force in a region of Georgia and has made several observations that closely parallel those in Newby's study of English farm workers.

The descriptive component of Barlett's study revealed data that contradict many widely held assumptions about full-time hired agricultural laborers in the South. Only about 60 percent were black, and, given their average age of 35, most were neither tied to one farm for a long time nor spent the majority of their careers as farm workers.

Similar to Newby's sample of farm workers, 82 percent had connections to agriculture before beginning work as a hired farm worker; 27 percent of their fathers were farm owners, 23 percent sharecroppers, and 32 percent farm workers. Barlett, however, found that these workers tended to have low educational backgrounds (though no lower than national averages for farm workers), work very long hours, and earn low wages, consistent with observations in other regions of the country.

Most farm workers, however, had few or no brothers or sisters in farm work, suggesting that they are less "trapped" in farm work than usually supposed. Barlett's analysis of the attitudes of farm workers toward their jobs was also similar to Newby's.

Newby reported that workers on large capitalist farms in England had much more positive attitudes toward their work than did industrial workers. Barlett reports very similar job satisfaction data for the sample of Georgia farm workers. Barlett also found that these Georgia farm workers were similar to Newby's sample in that they tended to have personalistic ties to their employers — frequently referring to their relationships with employers in family-like terms — and generally had a strong identification with the farm, the employer, and fellow workers.

Perry utilized quantitative analysis of census and USDA data on hired labour to explore a major anomaly in the conceptualization of hired farm workers. On one hand, Perry notes that many observers see hired labour in agriculture as low-opportunity-cost, "salvage" labour and that technological change is reducing the need for such workers in agriculture.

On the other hand, Perry cites data, such as those referred to earlier, indicating a trend for an increasing proportion of hired farm workers to be employed full-time and for the share of hired labour in total labour demand on farms to increase at the expense of family labour.

He has argued for a reconceptualization of the role of hired labour in agriculture. Following Braverman, Perry suggests that the use of hired labour should be viewed from the vantage point of the rationalization of U.S. agriculture, which causes farmers to adjust their use of labour, technology, land, and other resources in order to produce a profitable set of commodities with manageable risk.

Perry also follows Braverman in arguing that the role of technological change in affecting the use of labour will tend to be contradictory, given the imperatives of the rationalization process. On one hand, technological change may result in the simplification and deskilling of work, which encourages the use of cheap, unskilled wage labour (as opposed to family labour and more highly skilled hired workers). On the other hand, technological change may tend to create greater complexity in the labour process, which will encourage the use of more highly skilled labour such as farm family members and permanent workers. Perry's analysis of state-level census data suggests that the paradox of displacement of "salvage" labour by mechanization at the same time that hired agricultural labour is increasing in importance in U.S. agriculture can be accounted for by these two simultaneous and contradictory trends resulting from rationalization and technological change.

Index

G

I

L

M

N

O

P

R

S

T

U

W